Feminist Technology

WOMEN, GENDER, AND TECHNOLOGY

Series Editors

Mary Frank Fox
Georgia Institute of Techology

Deborah G. Johnson
University of Virginia

Sue V. Rosser
Georgia Institute of Technology

A LIST OF BOOKS IN THIS SERIES
APPEARS AT THE END OF THIS VOLUME.

FEMINIST TECHNOLOGY

Edited by Linda L. Layne,
Sharra L. Vostral,
and Kate Boyer

UNIVERSITY OF ILLINOIS PRESS
URBANA, CHICAGO,
AND SPRINGFIELD

Library of Congress Cataloging-in-Publication Data
Feminist technology / edited by Linda L. Layne,
Sharra L. Vostral, and Kate Boyer.
p.   cm. — (Women, gender, and technology)
Includes bibliographical references and index.
ISBN 978-0-252-03532-6 (cloth : alk. paper) —
ISBN 978-0-252-07720-3 (pbk. : alk. paper)
1. Feminist theory.
2. Women—Effect of technological innovations on.
3. Technology and civilization.
I. Layne, Linda L. II. Vostral, Sharra Louise, 1968–
III. Boyer, Kate.
HQ1190.F46315   2010
305.4201—dc22   2009045141

For feminists, past, present, and future

Contents

Preface

LINDA L. LAYNE

THIS SCHOLARLY EXPLORATION of and programmatic quest for feminist technology took root in the Science and Technology Studies (STS) Department at Rensselaer Polytechnic Institute (RPI). It makes sense that it should do so. This department is the home of Langdon Winner, who famously introduced the idea that "artifacts have politics" and who has done pioneering work on the politics of design, and is one of the cofounders of the department's Center for Cultural Design. His work is at the base of our exploration of the gender politics of artifacts and our belief that those politics can be designed into the materiality of things. "Just as surely as the rules, roles, and customs of everyday life and the laws and regulations of government, technological design is a place where some basic decisions are made about identities and relationships, power and status, life chances and limits upon these chances. In short, design is a profoundly political domain" (Winner 2002, 1).[1] Hence, we recognize technological design as an important site for feminist intervention.

In addition, this STS Department is known for its activist orientation. "Faculty and students pursue studies of power, gender, race, colonialism and the interactions between research and activism" (www.rpi.edu/sts). Many members of the department not only study social movements but also are active contributors to them.

It is also one of the departments from which the innovative, interdisciplinary Product Design and Innovation (PDI) undergraduate program was born. Brainchild of John Schumacher, a philosopher in the STS Department; Frances Bronet from the School of Architecture (one of the contributors to this volume); and Gary Gabriele, a mechanical engineer, PDI was launched in 1998.[2] The program emphasizes social awareness: "We aren't developing designers who can make a better widget for the sake of making one. We're developing designers who can go out into our increasingly interrelated global community and identify a need—not a want—study it, understand it, and address it" (Steiner quoted in Cleveland 2007/2008, 27). It is this emphasis that earned the program recognition in Business Week as one of the sixty "most forward-thinking design schools in the world" (October 15, 2007).

Two of the coeditors of this volume taught in the PDI program: Layne taught Studio 1 with two different interdisciplinary teams of architects, engineers, and artists, and Boyer taught Design, Culture and Society for the PDI program and regularly taught Design, History and Society, as well as Cities/Lands, a course that focused on urban design in contemporary and historical contexts.[3] Vostral came to design through her experience teaching Gender, Science and Technology and other elective courses to Rensselaer's engineering-heavy student body. She found that the best way to reach these students was through objects.

The larger setting of the university is also relevant. RPI has had an applied, if not activist, thrust since it was founded in 1824 by Stephen van Rensselaer, whose goal was to train people who would "be highly useful to the community" through their application of knowledge to the "business of living" (Rensselaer 2008a). Since the appointment in 1999 of its first female president, the university has adopted "Why not change the world?" as its slogan. In 2006 a vice provost for entrepreneurship was appointed and an E*ntrepreneurship initiative, which has a wide scope, was adopted. RPI is now trying to stimulate commercial, scientific, technological, social, and cultural innovation by integrating entrepreneurship throughout the curriculum. Its "Change the World Challenge" for students "awards $1,000 cash prizes to innovative ideas and inventions with the potential to make the world a better place" (Rensselaer 2008b).

In 2008, several of the prize-winning products were designed by female students, including a handheld device that enables early detection of skin cancer and stores pictures of suspicious spots on the skin so that they can be checked at a later date to see if they have changed, and an antibiotic bandage made of honey (a natural antiseptic), rice, and wax paper for use in parts of the world where individuals either cannot afford or do not have access to proper bandages (Rensselaer 2008b).

Another feature of Rensselaer is that it remains male-dominated. Although women were not formally forbidden from attending, with the exception of a few math students recruited in the early years for a special course, women did not start coming to RPI until 1942 when some were admitted "to replace men called to war" (Rensselaer 2007). Their numbers declined after the war. RPI began to recruit female students in the 1960s but despite efforts to increase their enrollment, the percentage did not reach 30 percent until the entering class of 2007. As for faculty, women represent about 18 percent of tenure-track positions (Berkley, Geisler, Layne, and Kaminski n.d.).[4] The number and proportion of women faculty decrease with each rank, and they have been grossly underrepresented in leadership positions.[5]

It is in this context that the STS Department has fostered collaborative, multi-disciplinary exploration of feminist issues with a particular focus on the intersection of gender and technology. Shirley Gorenstein was a founding member of the STS Department and its first chair (1983–95). Her training in archaeology has helped us keep in focus the importance of the materiality of technology. Deborah Johnson,[6] a philosopher who specializes in computer and engineering ethics, was also a founding member and chaired the department from 1995 to 1998. Together with Frances Bronet, a leader in the teaching of inderdisciplinary design who left RPI to become dean of the School of Architecture and Allied Arts at the University of Oregon, and another former RPI faculty member, Carol Collatrella, a scholar of comparative literature (now at Georgia Tech), we have worked in various combinations over the years to try to further our understanding of how gender, science, and technology enable and constrain women's lives.[7] Several volumes have resulted, including Gorenstein's 1996 volume Material Culture and her 2000 volume Gender and Work, and the Fox, Johnson, and Rosser 2006 volume Women, Gender and Technology. Feminist Technology is the latest in this series of collaborative efforts to advance our thinking on these issues.

NOTES

1. At a talk I gave on the book in 2008 to the Center for Interdisciplinary Research on the Family at Cambridge, a member of the audience suggested our title was flawed because, as she saw it, the problem was not in the technology but how any given technology is used. In other words, she adopted a classic "technologies are neutral" stance.

2. This started as a dual-degree program, but in 2007 it added the Design, Innovation and Society Bachelor of Science degree.

3. Our focus on design distinguishes this volume from others. With very few exceptions, design has received little attention in gender and technology studies. The most notable exception is the work done by Rothschild. She ended her essay in Machina Ex Dea (1983, 90) with the call to "discover how technology designed for the home . . . can be fashioned to benefit all women." Her 1999 collection Design and Feminism follows up on this call by bringing forward the work of feminist architects, urban planners, and graphic and industrial designers with an emphasis on the design of "space and places."

4. For more on the history of women at RPI, see Berkley, Geisler, Layne, and Kaminski (n.d.).

5. An NSF Advance Grant for institutional transformation was awarded to RPI to try to correct these problems.

6. In addition, Linnda Caporael has written an important piece on gender and tinkering and taught several courses on design including PDI Studio 1, Design Culture and Society,

and Psychology, Culture and Design. Nancy Campbell has contributed to feminist science studies through her work on the gendered politics of addiction.

7. Ellen Esrock, in RPI's Department of Language, Literature and Communication, has also contributed to these discussions, by, among other things, reminding us how and why "appearance" matters. She also contributed to Gorenstein's (2000) monograph on gender and work in public and private spheres.

References Cited

Berkley, Robyn, Cheryl Geisler, Linda Layne, and Deborah Kaminski. N.d. The Effects of Gender on the Promotion to Full Professor: A Case Study of a Technological University. Unpublished manuscript.

Cleveland, Amber. 2007/2008. "Innovation Education: Changing the Way Design Impacts the World." *Rensselaer Alumni Magazine* (Winter): 24–29.

Fox, Mary Frank, Deborah G. Johnson, and Sue Rosser, eds. 2006. *Women, Gender, and Technology.* Champaign: University of Illinois Press.

Gorenstein, Shirley. 1996. *Research in Science and Technology Studies: Material Culture.* Knowledge and Society Series. Stamford, Conn.: JAI Press.

———. 2000. *Research in Science and Technology Studies: Gender and Work.* Knowledge and Society Series. Stamford, Conn.: JAI Press.

Rensselaer Polytechnic Institute. 2007. Women at Rensselaer. www.lib.rpi.edu/Archives/ gallery/women. (Accessed September 3, 2008.)

———. 2008a. Rensselaer's History. http://www.rpi.edu/about/history.html. (Accessed September 3, 2008.)

———. 2008b. E*ntrepreneurship at Rensselaer. http://www.rpi.edu/about/entrepreneurship .html. (Accessed September 3, 2008.)

Rothschild, Joan. 1983. *Machina Ex Dea: Feminist Perspectives on Technology.* New York: Pergamon.

———. 1999. *Design and Feminism: Re-Visioning Spaces, Places, and Everyday Things.* New Brunswick, N.J.: Rutgers University Press.Winner, Langdon. 2002. Gender Politics and Technological Design. Paper presented at the Women's Studies Center, Colgate University.

Winner, Langdon, 2002. Gender Politics and Technological Design. Paper presented at the Women's Studies Center, Colgate University, October 22.

Acknowledgments

THIS EDITED VOLUME emerged from lively exchanges between valued colleagues schooled in different intellectual traditions and disciplines. Credit is due to these formal and informal collaborators and to the many feminists who inspired us to not only ask critical questions, but also propose solutions to our queries. The editors are grateful for our time together as colleagues at Rensselaer Polytechnic Institute, and for the opportunities posed by affiliations with new departments and the challenges of teaching at new institutions, which further strengthened this book.

Many thanks to the support staff at Rensselaer, Anne Borreo and Sara Feathers, for handling much of the document processing as well as international mailings required of the manuscript. Also, we greatly appreciate the editorial staff at the University of Illinois Press for fielding our many queries, with a special cheer going out to Joan Catapano for supporting the publication of feminist scholarship.

Khudija Mittu, a graduate student studying new reproductive technologies in the STS department at RPI, did a fabulous job creating a first-rate index.

Feminist Technology

Introduction

LINDA L. LAYNE

"FEMINIST TECHNOLOGY"—what a powerful and provocative concept; both immediately comprehensible and puzzling, simple on the face of it, yet complex. Both elements of this term are familiar, part of common language,[1] yet combined in this way they form something new. In 2007 I began a talk to the Women's Studies Department at Union College by asking the audience of mostly female undergraduates if they had ever heard the term, knowing full well that they had not. To my amazement, nearly three-quarters of the audience raised their hands and nodded, yes, they knew the term. I have had this experience with colleagues in the gender and technology field too. The reason for this, I believe, is that it is a term whose time has come.

The earliest use of "feminist technology" I have found is in a 1983 essay by Cor-lann Gee Bush. She does not define the term, which appears in her final sentence, but simply asserts that "A feminist technology should, indeed, be something else again" (1983, 168). She does give a good example earlier in the essay when she discusses how feminists were able to "unthink rape as a crime of passion and rethink it as a crime of violence, insights which led to the establishment of rape crisis and victim advocacy services. But a good feminist shelter home crisis service is something else again: it is a place where women are responsible for the safety and security of other women, where women teach self-defense and self-esteem to each other" (Bush 1983, 151).[2]

The term "feminist technology" did not catch on though, and an explanation for this might be found in the linked history of the term "feminist," which has had a spotty presence in what is commonly, and tellingly, known as "gender and technology studies." In the course of working on our book, we reviewed much of the gender and technology literature. As we anticipated, we found many insightful explanations and illustrations of how gender at once shapes and is shaped by science and technology but were struck by the fact that a feminist agenda (i.e., what we can/should do about these configurations) was strangely muted, even in those rare volumes that contain feminism in the title.[3]

In 1989 Joan Rothschild traced the change from sex to gender in the history of technology. She observed that "when feminist perspectives began to be applied to technology studies . . . the body of work was known as 'women and technology'" but that by the late 1980s, the term "gender and technology" was more frequently used. At first I thought our volume marked a third phase in the development of this field (from women/sex, to gender, to feminism) but the story is more complex. Feminism has been present throughout the development of this subfield but was more evident in pioneering works.[4] In the 1990s, "feminism" made fewer appearances,[5] perhaps in response to the backlash against feminism and/or the institutionalization of the subfield. In her foreword to a 2001 volume that boldly places "feminism" in the title,[6] Stimpson observes that many scholars avoid using "the F-word" and keep their feminism "covert" because they fear that since feminism is a political agenda, it might undermine the credibility (scientific validity) of their scholarship (2001, ix).

Johnson (this volume) also notes the subterranean position feminism tends to occupy in this literature. "Lurking in the background of much of the literature on gender and technology is an interest in social change, social change that will improve the circumstances of women and create more equitable gender relations." Both the term "feminist technology" and our eponymous volume bring this agenda to the fore.

I first used the term in 2004 when I wrote "The Home Pregnancy Test: A Feminist Technology?" for presentation at the annual meeting of the Society for Social Studies of Science in Paris.[7] After years of probing the relationship of gender, science, and technology, and working with young designers-in-training at Rensselaer Polytechnic Institute (RPI) to inculcate "the social" into every step of the design process, I was struck by the idea of "feminist technology" as a design goal. Why not encourage technological innovations that would enhance women's lives? This is our ambition.

Throughout this project we have found useful analogies with the environmental movement and the disability rights movement. Designing for differently abled people has become common in undergraduate engineering projects, and environmental concerns have permeated design curricula and popular consciousness. Our aim is to make feminism as pervasive in engineering, product, and architectural design curricula and practice as these other sociopolitical agendas.

Once we agreed to make "feminist technologies" a goal, we started with a series of definitional questions: Do any exist? If so, what makes a technology feminist? What criteria must feminist technology fulfill? Indeed, initially our intellectual project was framed as a question: "Feminist Technology?"

"Feminism" is a political movement that seeks to empower women. "Feminist" is a noun, referring to a person who works to empower women, and an adjective, which can be used to modify other nouns; for example, theory, organization, policy, or, in our case, technology. I define "technology" as tools plus knowledge that enhance and extend our human capacities.[8] Hence, a working definition of "feminist technology" would be those tools plus knowledge that enhance women's ability to develop, expand, and express their capacities.

Feminist technologies must "empower women," but we soon realized this was not as simple a criterion as it first seemed. Women are not the same. Must a feminist technology empower all women? Most women? What if a technology empowers some women and disempowers others? Furthermore, to "empower," a technology must give women more power over their lives. But lives are complex and we know that the introduction of new technologies always entails multiple consequences, some of which may be contradictory, many of which will be unintended. Given this, in order to determine whether a technology is empowering (for women, or particular groups of women, or some individual women), one would have to consider *all* the consequences of using (or not using) a technology. What benefits does this new technology bring? What costs? Then one would have to determine whether the costs outweigh the benefits, or vice versa. Should certain types of costs/benefits be weighted more than other types? For instance, in the case of a new reproductive technology, should physical costs/benefits count more than psychological or emotional ones? Who should decide: individuals, professionals, the state?

Next we invited people to think with us about these questions using their own case material and disciplinary backgrounds. In 2005, feminist scholars trained in cultural anthropology, archaeology, history, philosophy, geography, women's studies, architectural design and pedagogy, comparative literature, and art history presented their work at panels we organized at the Social Studies of Science Society (4S) in Pasadena and the Society for Literature, Science, and Art meeting (SLSA) in Chicago.[9] Some of the participants developed their case studies for this book. In each case, the product or products we studied are ones not only designed for and used by women, but also touted as liberating for women.[10] As it happened, all of these products (menstrual-suppressing birth control pills, home pregnancy tests, tampons, breast pumps, Norplant, anti-fertility vaccines, and microbicides) are intended to help women control their distinctly female reproductive systems. As such, they qualify as "feminine technologies," which McGaw defines as "technologies associated with women by virtue of their biology" (McGaw 2003/1996, 15). By no means do we equate "feminine technologies" with "feminist" ones. However, as we began the process of querying feminist technologies, we found these technologies to be "good to think" with. They are in some ways, the simplest test cases and yet, as we quickly found, not simple at all.

By the time we presented our papers in 2005, we had come to some preliminary conclusions. At this point we knew that marketing a product as liberatory for women does not make it so. Efforts to use the language of women's liberation to sell products, including ones that are harmful to women, are not new. Faludi (2006, xiv) gives the example of a 1929 Easter "Freedom March" down Fifth Avenue "organized by a prominent ad man . . . to honor suffrage—by encouraging women to smoke. The American Tobacco Company's publicist persuaded 'a leading feminist' to head up the procession of women, who were all puffing on their 'torches of freedom.'" Johnson (this volume) describes a similar marketing strategy for Virginia Slims cigarettes. More recently, Hanes apparently "persuaded a NOW official to endorse its 'liberating' pantyhose" (Faludi 2006, xiv). These might be considered examples of "feminist washing," akin to "green washing," the common practice of marketing an unsafe product as environmentally friendly in order to attract consumers with environmental concerns. Each of our case studies brings a critical lens to the way the rhetoric of feminism has been used to promote consumer products.[11]

Another conclusion we reached at this point was that feminizing an existing technology did not make it feminist, but often just the opposite. In response to Wosk's presentation on new electronics such as pink, rhinestone-encrusted cell phones being designed for and marketed to women, Gorenstein observed, "while recognizing the utilitarian benefit of the phone to women, making them more mobile and less housebound," such phones appear to be an example of manufacturers realizing that they had designed their product in the first place for men, not for people. They now had to alter the design so that the product, or a version of the product, was feminine. The next question they had to ask is how are women different from men? A common answer—the most common—is that women have a different aesthetic sense from men.

The changes they made based on their understanding of "women's aesthetic" ("brighter colors, curved lines, and surface adornment") raised for Gorenstein "the question of whether these were actually women's choices or men's choices for women."[12] Gorenstein concluded that

> if *Sex and the City* is right and the powder-compact color-drenched jewel-adorned cell phone is what women want and reinforce women's goals and interests, then I would venture to say that the manufacturers are producing a feminist technology. However, if this technology reinforces a gender system in which women's aesthetic has been defined by men to suit men's interests and

are antithetical to women's interest and goals, then the designers and manu- facturers have certainly not produced a feminist technology, and indeed may have produced an anti-feminist technology. (Gorenstein 2005)[13]

Much of the innovation recently directed toward women has been of this kind: based on a "shrink it and pink it" approach that, as cultural anthropologist Gen- evieve Bell of Intel observes, is "profoundly misguided" (National Public Radio 2008b). The Gadgettes podcast on Cnet.com has a "pink watch" segment that focuses on gadgets that are "gratuitously pink."[14] "Bubble-gum pink stun guns" would certainly be a candidate. According to a saleswoman for the product, these guns are "all about girl power" because they "fit in purses." They are being sold at "Tupperware-esque 'Taser parties'" even though "there's a lot of evidence that stun guns could be used against the owner" (S. W. 2008, 116).

Another notable example of this type of product is a Samsung cell phone that has a "pink exterior, built-in shopping list, fat calculator, . . . conception calendar, favorite fragrance list and biorhythms." In addition, according to Jenny Goo- dridge, the marketing manager for this product, "You can put in your birth date and it will tell you if you are intelligent, attractive or emotionally stable. You can't be all three." As one commentator observed, this "begs the question, do women really need an ovulation calendar on their mobile?" (Manktelow 2005).[15]

More promising technological innovations are those that enable women to enter and do their best in professions that had traditionally been held by men and for which the necessary equipment had been designed to fit men's bodies. Johnson (this volume) discusses the case of the fighter pilot cockpit that was redesigned in order to allow for the integration of women into this part of the military. Other examples include construction, auto repair, firefighting, and law enforcement in which "poorly proportioned tools, uniforms and other gear can affect performance and safety" (Melendez 2006). For instance, traditionally designed police duty belts, from which the gun, handcuffs, and other gear hang, do not fit comfortably on a woman's hips and may cause bruising, and firefighting gear was designed for "fairly tall, well-built, muscular men" (Melendez 2006). Designs that address women's toilet needs like the race-car driver's suit for women described in Bronet and Layne (this volume) also facilitate women's entrée and success into tradition- ally male-dominated professions and leisure pursuits.[16] Medical technologies that take into account anatomical differences are also valuable improvements. For example, Stryker Orthopedics developed a special line of knee replacements for women because women's knees tend to be smaller than men's and proportionally narrower (Melendez 2006).

Yet we also realized that designing for "equality" is not an adequate goal. In

Gorenstein's comments on the SLSA papers, she made this point, "Suppose men are using a factory technology that is dangerous. [Should] women . . . strive to earn the right to use that technology in order to be equal to men?" The answer is clearly "no" (Gorenstein 2005).

We also refined our understanding of the limitations of a definition based on "improving the conditions of women" without linking this with inequities based on gender (Johnson, this volume). A good illustration of this can be found in research on barriers to advancement of women faculty at RPI. Geisler et al. (2007) created a new information technology, the "13+ Club Index," which is an easy-to-use tool by which patterns of inequality in men's and women's rate of promotion can be determined and displayed. In 2001, the index revealed that many professors at RPI (both men and women) were still associate professors more than thirteen years past their terminal degree (the point at which one would normally expect to be promoted to full professor) but that women were more than twice as likely as men to be stuck in rank. The results of using this technology were dramatic. Five of the eleven women in the 13+ Club were promoted to full professor (two of whom had been previously denied), an unprecedented number and rate of promotion for women at RPI. Yet a follow-up study in 2004 found that although the index had benefited some women, it had also benefited some men (Geisler et al. 2007).[17] This, combined with other aspects of the hiring and promoting system, like the pattern of senior hires going to men (35 percent of the male hires between 2001–4 were at the full professor rank; none of the female hires were), meant that the inequality in promotion rates remained virtually the same. It is worth noting that when women faculty representatives met with the provost in 2001, we "discussed changes in policy and procedure that could improve the advancement picture *for both men and women*" (Geisler et al. 2007; emphasis added) because we assumed that this innovative technology would more readily be accepted if it could be shown to benefit men too. The pitfalls of such a strategy are apparent.

To summarize, marketing claims do not make a product feminist, nor does the dressing up of an existing technology to conform to a male-defined "feminine aesthetic." More promising are innovations that adapt a technology designed for and used by men so that women may also use it. However, in these cases, the overall effects on women must be considered. Even a technology that improves things for some women may not qualify as feminist if it does so in a way that perpetuates a gender gap.[18]

As we began to find answers to some of our questions, we discovered new ones that needed to be considered. Are feminist technologies simply or necessarily artifacts "designed by women, for women"? If a technology is feminist, how did it get that way? Is the feminism in the design process, in the thing itself, in the way it is marketed, or in the way it is used by women and/or by men? If one were to set out to design feminist technologies, how would one go about it? Might a technology be considered feminist in light of one type of feminism and not feminist, or even antifeminist, according to another? If one engages in feminist design practices, does one necessarily end up with a feminist design? Can one end up with feminist technologies without engaging in a feminist design process? What are the consequences if one begins, explicitly, with a feminist agenda for product design and innovation? What can we learn from failures? Some of these questions are still not answered to our satisfaction, but enough have been over the course of working on this project that at this time we feel confident enough to drop the question mark from our title. We now know that feminist technologies are not necessarily "designed by women, for women." Neither the sex of the designers nor of the intended users is definitive.

One candidate for a feminist technology designed to be used by men is the male pill (Oudshoorn 2003). Second-wave feminists advocated the development of new male contraceptives including a male pill once the health risks of the female oral contraceptive became known, calling for men to share both the risks and responsibilities of family planning. The male pill is also an example of what Oudshoorn calls a "gender bender" because it challenges the hegemonic gender system, in this case by challenging "hegemonic masculinity" (2003, 16).[19]

Another technology designed for male use (one that responds to but does not challenge hegemonic masculinity) is the electronic anklet designed to protect women from intimate partner violence. Such electronic monitoring systems have been put into use in several counties in the United States, including Brooklyn, New York; Shelby County, Tennessee; San Bernardino County, California; and Pitt County, North Carolina, in response to the high rate of domestic homicides perpetrated by men who had been arrested for domestic violence but were free on bond (Weigl 2006; see also Herszenhorn 1999). The time between arrest and trial is known as "the riskiest period for victims."[20] The satellite-based monitoring system alerts the police officers if these men go anywhere near their victims' homes or workplaces."[21] Another version might automatically deliver an electrical shock to the abuser if he gets too close or allow the woman to activate the shock to protect herself).[22]

Candidates for feminist technologies designed by men for women include the vibrating tampon (Vostral, this volume) invented by a man to help alleviate his wife's painful menstrual cramps. Another is a television remote control override panel called "Stop It" designed by an engineering consultant for his wife so that she could "freeze the channel" and prevent his channel-surfing (Bellamy and Walker 1996, 125).[23]

Although feminist technologies need not be designed by women—that is, the sex of the designer is not a requirement—it is more likely that feminist technologies will be designed by women because the life experience of a designer informs every aspect of design, including problem identification and selection. This dynamic is evident in the work of the women entrepreneurs I have encountered through my research on pregnancy loss. For example, Sandy Maclean of Wisconsin developed a product after having a traumatic miscarriage at home for which she had been inadequately prepared. As a result, she came up with the idea of a home miscarriage kit (patent pending) that includes heavy sanitary pads, disposable bed pads, an informational brochure with contact information, and a container in which to put the embryo or fetus, which she and her business partner hope to distribute through clinics (Layne and Bailey 2006).

Another example of this can be found in the Web-based company Our Hope Place, started by Laura Racanelli and Sharon Stenger following their miscarriages. The women had become friends while completing degrees in mechanical and process engineering at RPI. When Laura had a miscarriage, the jade bead bracelet her mother gave her helped her while she was grieving and during her anxiety-ridden subsequent pregnancy. When her friend Sharon had a miscarriage, Laura passed the bracelet on to her. Because there are no cultural scripts for how to respond to a pregnancy loss, friends and family are unsure what to say and as a result often do not even acknowledge the loss. In response to this cultural lacuna, Our Hope Place is designed to help friends support their friends when they have a loss; they offer advice on what to say and a selection of friendship gifts.

In both of these cases, the products are not new; the innovation is in deploying existing products to meet the unrecognized needs of women who have pregnancy losses. In a similar vein, my 2003 book (*Motherhood Lost*) and my television series, *Motherhood Lost: Conversations,* can be seen as educational technologies designed to help address the inadequate public understanding of pregnancy loss.

I also encountered many new products that were being developed by women entrepreneurs as a result of their personal experiences at the International Doula Conference held in Vancouver in 2008. The exhibition hall was filled with women who had started companies, usually in partnership with a female friend, that marketed products to pregnant women or new mothers, products that they wished they would have had during their own pregnancies or as new mothers. One such product is PumpEase, a patent-pending, hands-free pumping support developed by a Canadian mother with a degree in fashion design and technology.[24] Wendy Armbruster Bell came up with the idea of a device that would enable her to do other things while pumping because she found pumping so boring. As she explains on her Web site, "After . . . searching the internet, only to find ugly, expensive and fiddly pumping apparatus, I mocked-up a pumping support . . . which was not only functional and user-friendly, but pretty darn funky too! . . . After more than 400 pumping sessions, I [knew] what a pumping support should offer a nursing mom—the ability to free her hands for other tasks such as reading, talking on the phone, using a PC or tending to her baby."

Another product on display at the conference was an ergonomic, environmentally friendly baby carrier made out of three pieces of stretchy bamboo fabric that can be worn six different ways. It was designed by two Canadian women: Alison Cross, a mom who couldn't find a sling that was easy and comfortable to use, and Paula Violi, who was trained in industrial design. The Baby Buddha

A retro-style, hands-free, double breast pump support.
Photo provided by PumpEase.

TUBE SLING (A)	TUBE SLING (B)	BELLY WRAP

position	age	benefits
womb with a view	newborn to toddler + nursing	for sleeping and nursing
together baby	4 months to toddler	for babies who prefer to be in an upright position
the tourist	4 months to toddler	for babies who prefer to look out
baby buddha	from 4 months	when you want to travel light and only use one sling
hip baby	4 months to toddler	a perfect hands free carry for toddlers
piggy back	toddlers	for children who are a little heavier

A three-piece stretchy baby carrier for newborns to toddlers that "gives you the freedom to go anywhere and do anything." Image provided by Baby Buddha.

Web site states their philosophy that "parent-led design vetted by an industrial designer will ensure parents get products they really need."

Other products included several models of nursing pillows that allow women to nurse while sitting up without hunching over and straining their backs and shoulders; an elastic prenatal cradle that cradles a pregnant woman's abdomen and back; organic cotton, washable, postpartum sanitary pads; a trendy line of clothing for birthing designed by a woman who is a "doula and mother of three"; and a company whose aim is to help these other companies succeed.[25]

More surprising than the fact that feminist technologies need not be designed by women is the realization that they need not be designed by feminists. Indeed, some technologies that benefit women have sexist origins. One such example is the now-iconic dial pack for birth control pills that was invented in 1964 by product engineer David Wagner "to help his wife remember to take her pills and to overcome the marital rows that emerged around her taking the pill" (Sanabria 2008). Of course, it is in women's best interest that their birth control method be effective and the pill is only effective if taken regularly, as directed. So even though the ring pack allowed the male inventor (and potentially other men) to engage in surveillance to see for himself whether his sexual partner was being "medically compliant," the product has enhanced women's ability to control their fertility. Furthermore, if and when the male pill comes into use, women may find the ring dispenser useful in supervising their male partner's medical compliance.[26]

The reason that feminist technologies can be designed by sexists is that our criteria are consequentialist. What matters is their effect, not their intended effect. This is also evident in the fact that some feminist technologies are feminist by accident; that is, the benefit to women is an unintended consequence. Take, for example, the curb cuts, ramps, and larger specially designated restroom stalls that became pervasive in the United States following the American with Disabilities Act of 1990. These innovations were instituted to make public spaces accessible to people with motor impairments (e.g., those who use wheelchairs or walkers)

This still from the film *Serious Undertakings* (Helen Grace 1982; photographer: Sandy Edwards) appeared as part of a three-part series in Wajcman's *Feminism Confronts Technology* (1991) and illustrates the need for feminist design. Reproduced with permission.

Wheelchair access   **Pushchair route**

Pictograms used on London public transport. Space
is reserved for both wheelchairs and baby buggies on
London buses. Note that both images use unisex figures.
Reproduced with permission of Transport for London.

but have been a great boon to those who use strollers. The complementarity of
spatial and surface needs of stroller and wheelchair users is now sometimes explicitly acknowledged, as in the United Kingdom where open space is reserved
on buses to accommodate both strollers and wheelchairs.

However, although innovations that are designed for sexist reasons or are
designed without women even in mind may benefit women, these are crumbs
compared with what we could have if women were demanding that the world be
built to better meet their needs (why hadn't women asked for curb cuts, ramps,
and family-size toilet stalls in the first place?) and if technologies were being
designed by feminists.

Take, for example, the Tampax tampon that, as Vostral explains, was invented
by a sexist.

> Earle Cleveland Haas . . . admitted that he conceived of the design because
> he "just got tired of women wearing those damned old rags." He also pitied
> women during their menses . . . [When asked] how he became interested in
> designing a tampon he replied, "Well, I suppose I thought of the poor women
> that got in that mess every month, you know. It was very disagreeable to them
> and I thought, well, surely there is a better way to take care of that." (Vostral
> 2008, 77)

To the extent that tampons are a feminist technology (and Vostral 2008 develops
arguments both for and against this), the sexist attitudes of the inventor seem
to be irrelevant.

On the other hand, as Gorenstein observes,

> A designer who believes the design problem is to contain menstrual blood
> will invent a different technology from the designer who thinks the design
> problem is first about women's well-being and second about catamenia. For

example, in choosing between a plastic or natural batting for the tampon, the feminist designer . . . weighs information about contact dermatitis caused by plastic differently from the designer looking for the most absorbent material for the device" (Gorenstein 2007)

Gorenstein points here to the importance of a holistic view of women's well-being. In addition, Gorenstein argues that a feminist technology must both benefit *and* empower women

> because a technology that benefits women but does not empower them is paternalistic, leaving women in the hands of others whose goodwill must be courted. It places a potential obstacle between women and their better life. Empowering refers not only to, on balance, beneficial effects but also it is connected to the ability to take action as an individual and as a member of a group to achieve those beneficial effects. (Gorenstein 2005)

This is similar to the distinction made by Bush (1983) about types of rape crisis centers (i.e., between those that provide needed services to women, and those that train women to defend themselves so they will not need such services, and train women to provide the services to each other when needed).

This leads us to another key element of feminism. In 1968 Aileen Kraditor defined "feminism" as "the assertion of female autonomy" (quoted in Sklar 1998, 180). Writing of nineteenth-century feminists, Kraditor observes, "What the feminists have wanted has added up to something more fundamental than any specific set of rights or the sum total of all the rights men have had. This fundamental something can perhaps be designated by the term 'autonomy.' Whether a feminist's demand has been for all the rights men have had, or for some but not all the rights men have had, or for some men have *not* had, the grievance behind the demand has always seemed to be that women have been regarded not as people but as female relatives of people" (quoted in Sklar 1998, 180–81).

If we focus on technologies that enhance women's ability to assert their autonomy from men, the vibrator (Maines 1999) and assistive reproductive technologies like sperm banks and artificial insemination that have facilitated lesbian parenthood (Lewin 1993) and the ability of heterosexual women to choose to mother without the involvement of a man come to mind. In a review of the book *Knock Yourself Up* in the newsletter of Single Mothers by Choice (SMC), an organization established in the United States in 1981, one woman explains, "[I]n an effort to ensure our own security and happiness, we are changing the order in which we do things, often becoming the very person with whom we wanted to partner" (Tujak 2008, 8).

Should we reserve the appellation "feminist technology" only for those technologies that empower and do not merely benefit women? As eager as I am for technologies that empower women, I am loath to define "feminist technology" too narrowly; to shun innovations that would make things incrementally better for women, even if another innovation would be better.

Perhaps it would be useful to think of technologies or technological innovations as fitting onto a continuum in terms of feminism. Instead of a single term, we might want a series of terms that mark degrees of feminist achievement—for example, "minimally feminist technology" for those that improve things for women some degree from the status quo; "moderately feminist technology" for innovations that provide a substantial improvement for women over the status quo; and "radically or truly feminist technology" for innovations that adopt a holistic approach to women's lives and make changes that radically restructure arrangements in ways that will benefit women and substantially shift the balance of power between women and men.[27] This would permit one to acknowledge and embrace innovations that offer some benefit while not losing track of the goal of innovations that are optimal for women. It might also encourage more radical innovation. An analogy can be drawn with the ranking systems for "green" buildings like the Leadership in Energy and Environmental Design (LEED) Green Building Rating System, a nationally accepted certification program that ranks energy-efficient buildings. This initiative has, in turn, generated a profusion of other, more radical ways of conceiving of and ranking "energy-efficient," "sustainable," "regenerative," "eco-revelatory," "living" buildings.[28]

We have learned that coming to consensus on what constitutes a feminist technology is very difficult, and introducing a graded typology of feminist technologies would no doubt compound this. But we have also come to appreciate the generative

LEED crest on the Getty Center, Los Angeles. Photo by Linda Layne, 2008.

power of efforts to define feminist technology and the hope would be that such a gradation might likewise be constructive in this regard.

Indeed, one of the most important things we have learned is that we will not necessarily agree on whether any given technology should be considered feminist. This has been both one of the most frustrating, but ultimately valuable, lessons of working on this topic. Each time we came up with a technology that we felt qualified as "feminist," someone would invariably, and almost immediately, disagree. They would point out negative effects we had not considered, or suggest a better solution, or show us how we had misunderstood the problem and how if looked at in a different way, another solution would clearly make more sense.

One of my earliest experiences with this was in the form of comments by an anonymous reviewer to the suggestions I made for improving home pregnancy tests in an earlier draft of my chapter (this volume) on that subject. One of my suggestions was to create different tests for different users (one for those who want to be pregnant and one for those who do not). The reviewer noted that such need-specific tests would exacerbate the problem of privacy. S/he imagined a scene at the checkout counter: "I see you're buying the morning-after test. Are you planning to abort if it's positive?" or "Is that the two-week test you've got there? You do know that we have next-day tests if you're planning an abortion. The two-week tests are more appropriate if you've had earlier pregnancy loss. So is that the one for you?" Of course, s/he is right. Why had I not thought of that?

In addition, where I advocated that women spare themselves the pain of early losses by postponing the point at which they ascertain if they are pregnant, this reviewer pointed out that for some women the knowledge that they were pregnant, even if only for a few days or weeks, may be valued and they should not be deprived of that. In this case, I had thought of that—indeed, I have written with sympathy about this in my book (2003) on pregnancy loss—yet my own preference would have been to have been spared the experience of loss and I believe that some others, if given the choice, would share this preference.

Another example of a technology that seemed, at first, like a great example of a feminist technology is the cry room—a soundproof, glass-fronted room with "piped-in" sound found in some movie houses and houses of worship. These technologies allow people (female or male) who are responsible for the care of infants and toddlers to participate in these communal activities without disturbing others.[29] This architectural innovation was suggested to us as a feminist technology by a member of an audience at a talk I gave to art and architecture students at the University of Oregon because it allows mothers of young children to engage in social pursuits they would be excluded from otherwise. But others quickly pointed out that given the extent to which women are considered responsible for young children (the wife-to-husband ratio for child care in the United States is still close to five to one (Belkin 2008, 47), it would likely be women who are expected to separate themselves from other adults to use these facilities.[30] Even cry-room designers note that "feelings of exile can surface" when one "steps out of the spiritual community of a sanctuary" and they speculate that may be why many cry rooms "are seldom used" (Bott 2007). Bronet (personal communication) suggested that a better alternative is found in those Jewish temples where children are allowed to move between the women's and men's sections of the sanctuary during services and all attendees share the responsibility of caring for them.

Box pews like the ones found in the Old First Church, built in 1805 in Bennington, Vermont, were perhaps designed with similar concerns in mind. Parents and their young children occupied an enclosed space with barriers high enough to contain toddlers but low enough not to obscure the alter.[31] These pews, which look like family-sized playpens, while not solving the problem of noise, would at least keep toddlers from toddling off. But rather than seeing this as a benefit to women, the feminist friend with whom I discovered the pews saw them as a corral that immobilized women and children in order to indoctrinate them with patriarchy.

Box pews, Old First Church, Bennington, Vermont. Permission granted by photographer Nancy Andrews.

A related innovation was discovered by Boyer (personal communication) after moving to the United Kingdom. "The Big Scream" is a Parents and Babies Club at movie theaters that has special midday showings of films "during which the lights are left slightly up, the volume is turned down a little, and the babies can cry as much as they like." Some theaters also offer "a nappy changing facility, bottle warmers, and a safe area for pushchairs" (Despois 2005). The club is marketed as "the perfect opportunity to meet other mothers/fathers/carers with children of a similar age" and patrons are encouraged to stay after the film to "eat, drink and socialize."[32]

Eventually, we learned that rather than be discouraged by these exchanges, we should welcome them. Design is based on exactly this type of critique. Designers are encouraged to think creatively, to think proverbially outside the box, and, at the beginning, not to worry about whether the ideas are practical. As the design process progresses, many ideas are generated, presented, abandoned or revised, presented again, and so forth. An iterative process that involves continual self-critique and periodic critiques from peers and outside reviewers is fundamental to design. It is exactly this type of creative brainstorming, give and take, and experimentation that we want to encourage.[33]

Advertisement in the Cambridge Arts Cinema for their Parents and Babies Club. Photo by Linda Layne, June 9, 2008.

One of the reasons we will not always agree on what constitutes a feminist technology is because feminism is "a many splendor'd thing" (Meyer 1987, 389). The feminist approaches by which we evaluate technologies are diverse. A technology may appear feminist in light of one type of feminism and antifeminist through a different feminist lens. A good example is the menstrual-suppressing birth control pill, which Aegnst and Layne (this volume) show is considered feminist from the perspective of some feminist theories but rejected as antifeminist by others. Another example is found in Norplant and the antifertility vaccine described by Hardon (this volume). In both cases, radical feminists rejected these new forms of birth control as irredeemably antifeminist, while liberal feminists felt that with some modifications they might be valuable additions to the range of contraceptive choices available to women.

Other examples are the technologies of bottle feeding (baby bottles, formula, and/or breast pumps), which from a liberal feminist position are viewed positively because they "make it possible for men and women to share equally (though differently) the responsibilities of infant care" (Johnson, this volume). Yet from an essential feminist perspective, like that evident among proponents of attachment parenting, such technologies are considered antifeminist. As Faircloth reports from her research in the United Kingdom, "[S]ome women consider bottles (and strollers, dummies [pacifiers], cots, etc.) as 'mother replacement utensils'" that devalue the mothering role by encouraging women to separate from their babies and they resist what they believe is "pressure . . . to share the job of mothering" (Faircloth 2009).[34]

Bottle-feeding also illustrates the importance of cultural context. As Gorenstein points out,

> There are a number of contextual issues around the baby bottle which can be highlighted by the famous 1970s Nestlé case in which that company manufactured a powdered infant formula that was sent to poor third world countries. Women of these traditional societies began making the choice of using the baby bottle and the formula. They reasoned that the technology would allow them to work outside the home because non-mothers would be able to feed the baby. The problem was that since the formula required water and water was contaminated, women who used the formula were endangering their babies. The proposed solution by some feminist groups, particularly breastfeeding advocacy groups, was that the company should be boycotted (still in effect, by the way) until they stopped sending the product to these countries. The boycott did not demand that the company provide a canned or boxed liquid formula that didn't require water or refrigeration (easily done technologi-

cally) which would enable women and men to use the baby bottle safely. The boycott assumed that restoring breastfeeding as the exclusive way of feeding babies was the right goal. In this situation, the groups from non-traditional societies (where the baby bottle made breastfeeding optional) were reinforcing traditional society's historical gender system (where breastfeeding was done exclusively) while women from traditional societies were moving towards changing or subverting their current gender system. How do we answer the question, is the baby bottle and formula a feminist or anti-feminist technology? We can't answer that question without working through not only the complexity of the context of the situation in which the technology is set, but also the complexity of the situation of the activists. (Gorenstein 2005)

Another element of context that must be taken into consideration is the cultural specificity of patriarchy. If what feminists are doing is trying to free women from the constraints imposed by patriarchy, it is important to recognize that the patriarchal systems against which they rise up differ. This is one of the main lessons of Meyer's (1987) comparative study of *Sex and Power: The Rise of Women in America, Russia, Sweden, and Italy.*

A comparison of Wi-Fi technology used by women in the United States and Japan helps illustrate how such differences affect the potential and use of new technologies. Genevieve Bell, director of user experience at Intel, explains why American women led men in early adoption of Wi-Fi technology.[35]

Men and women have different lives and demands on their time. If you look at the data from the U.S. Department of Labor or data from the United Nations, one of the things that is really striking is the ways in which the demands on women's time haven't in fact changed in nearly fifty years. We spend as much time now doing housework, child rearing, and emotional work for those around [us] as we did fifty years ago and add onto that our presence in wage labor. (National Public Radio 2008b)

Indeed, the University of Wisconsin's National Survey of Families and Households shows the average ratio for housework in the United States is about two to one, with "the average wife" doing thirty-one hours of housework—for example, cooking, cleaning, yard work, and home repairs—a week "while the average husband does 14." In households where both partners have full-time paying jobs, "the wife does 28 hours of housework a week and the husband 16" and even if the woman works full time and the husband doesn't, the wife still does the "majority of the housework" (Belkin 2008, 46). Another recent study done at the University of Michigan Institute for Social Research found that when a man and woman marry, the woman does an average of seven more hours of housework a week and the man does about one hour less (Young 2008).[36]

This is the context in which middle-class American women adopted Wi-Fi technology. The ability to gain access to the Internet from a number of locations was recognized by American women as a useful tool to help them "juggle a lot of competing demands." For instance, they could "do the family banking while taking a kid to the dentist, to be at a kid's football game and still be able to do the family shopping" (National Public Radio 2008b).

In Japan women have also been early and avid adopters. Like their American counterparts, Japanese women are using Wi-Fi technology to negotiate culturally imposed patriarchal constraints on their lives; however, these constraints are different and hence the uses to which they put this technology are too. One of the main uses for "tens of thousands of married Japanese women" is speculative trading in currencies. The scale at which they are participating is such that they have "a noticeable impact" on global currency markets (Schultz 2008, 30). Like American women, they appreciate the way Wi-Fi helps them integrate other activities while fulfilling the traditional gender role of taking care of children. "Online trading can be done while a woman is at home taking care of kids, during a lunch break, or even in the middle of the night if she is a currency trader or trading stock on a foreign market" (Schultz 2008, 30). An important difference, though, is that there are many fewer opportunities for Japanese women to achieve financial independence. The gender wage gap is even greater in Japan than it is in the United States (women's wages average 65.5 percent of men's wages in Japan versus 78 percent in the United States) (Schultz 2008, 6), and employment opportunities for women are fewer. "The Japanese 'lifetime employment system' favors workers who can make long-term commitments to a company" and as such "is inherently biased against women who are also expected to commit to their families. Consequently, many women are excluded from the employment system" and personal investing provides a welcome alternative (Schultz 2008, 19). In addition, Japanese women are acutely aware of the need to provide for themselves in old age. The Japanese system of social security is in crisis as a result of the low birth rate and the high percentage of young workers who are not paying into the system (Schultz 2008, 20). All this helps explain the popularity of personal investing but not the special attraction to Japanese women of doing so via a cell phone. For this, we must look again to the particular contours of Japanese patriarchy. Wi-Fi technology enables housewives to trade secretly. "Some women use their own money, but others use their husband's money or joint earnings without his knowledge" (Schultz 2008). In addition to providing a degree of freedom within the confines of a patriarchal marriage system, the anonymity that online trading provides is also valued by Japanese women because it protects them from gender discrimination while investing. Unlike American female investors who tend to prefer dealing with a real person, Japanese women recognize that the anonymity of online investing means "gender [will] pose no constraints on earnings and capacity to succeed" (Schultz 2008, 20).

In both the American and Japanese cases, we can see why women adopt technologies that help them "lead their lives"; that is, to operate more fully within the constraints of the locally specific forms of patriarchy. But clearly technological fixes are not enough. Feminists must also work toward undoing patriarchy in all its forms. This means not only introducing new technologies, but changing technosocial systems (Johnson, this volume), including an equitable division of family labor and eliminating discrimination in the workplace, in private investing, and in public life.[37]

Social fixes like the French Parity Law that requires "French political parties to present equal numbers of male and female candidates" (Campbell 2003, 121) and the Gender Equality Duty passed in the United Kingdom in April 2007 are helpful. This new British law requires "public authorities to promote equality between men and women and eliminate unlawful sex discrimination" and is an improvement over U.S. sex discrimination laws in that instead of depending on individuals making complaints, it "places the legal responsibility on public sector organizations, authorities and institutions to demonstrate that they actively promote equality between men and women. The duty affects policy making, public services such as transport, and employment practices such as recruitment and flexible working" (Equality and Human Rights Commission 2008). One area in which this law is anticipated to have a positive impact for women is urban regeneration, an enterprise that involves "housing, transport, education, health, and crime," all of which affect women and men differently (Burgess 2008, 4). In England, as in the United States, "women in general have lower incomes, they are more likely to be carers," and they make up the vast majority "of lone parents" (Burgess 2008, 4), and yet have been historically underrepresented in decision making about urban renewal. A new initiative between Oxfam and the Women's Design Service is working with planners, officials, local residents, and local women's groups to implement the new Gender Equality law so as to "root out hidden and structural gender discrimination in regeneration programmes" (Burgess 2008, 4).

Thus, we recognize what Johnson (this volume) describes as the chicken-egg problem. Which came first, sexist society or sexist technologies, and, conversely, what should we seek to change first, society or technology, given that they shape and are shaped by each other? We also recognize that regardless of the intent, new technologies introduced into unjust social systems are apt to end up reinforcing the status quo. As Woodhouse (2008) puts it, "[N]ew technoscientific capacities introduced into an inequalitarian society will tend disproportionately to benefit the affluent and powerful." He advises that "those who seek to make the world fairer," should "oppose technoscientific research and innovation unless it aims disproportionately to assist" the underprivileged. Given all this, how, where, and when can we most effectively promote technologies that disproportionately benefit women?

This appears to be a particularly opportune time to champion feminist technologies. In the United States, women are expected to control one trillion dollars, or 60 percent of U.S. wealth, by 2010, according to research conducted by *Business Week* and Gallup. Women "purchase or influence the purchase of 80 percent of all consumer goods, including stocks, computers and automobiles" (http://www.microsoft.com) and they make 45 percent of consumer electronic purchases (National Public Radio 2008a). In an article entitled, "I Am Woman, Here Me Shop" published in *Business Week* in 2005, women are deemed "the 'apple of the marketer's eye' . . . thanks to their . . . purchasing power and decision-making authority. Working women ages of 24–54—of whom the U.S. has some 55 million—have emerged as a potent force in the marketplace, changing the way companies design, position, and sell their products." This is attributed in part to the fact that 27 percent of households are now headed by women, "a fourfold increase since 1950,"[38] and the fact that "in the past three decades, men's median income has barely budged—up just 0.6 percent—while women's has soared 63 percent. Some 30 percent of working women out-earn their husband" (Gogoi 2005).

Pat Schroeder, who was a U.S. congresswoman for twenty-four years, has a good idea about how to use women's buying power for feminist ends. She notes that in addition to making the majority of purchasing decisions within families, women also do so in corporations. "It is women who are making the hotel reservations, the airline reservations and pretty much everything else. We need a way to rank the major corporations . . . on whether they are good for women and let women purchasers know. . . . Does the company have women on the board of directors? Does it have good work and family policies? Does it have women in their top line management?" (Schroeder 2003, 88). A frequently updated Web site would make this information accessible so that women could choose companies "that have more progressive policies toward women" (Schroeder 2003, 88). Even if only 10 percent of women would change their buying habits, they would have a tremendous effect. "No one can stand to lose 10 percent of market share" (Schroeder 2003, 89).

Of course there is a risk in advocating liberation through consumer goods. Indeed, in the new preface Faludi wrote fifteen years after the publication of *Backlash*, she explains the failure of the feminist movement to achieve its potential via the metaphor of Ovid's story of Atalanta who finally lost a race against a man when her opponent rolled three golden apples in her path that she paused to scoop up. Like Atalanta, Faludi says, feminists got distracted from their goal by "the trinkets [that] are the bounty of a commercial culture, which has deployed the language of liberation as a new and powerful tool of subjugation" (2006, xiv). By the early years of the twenty-first century, Faludi (2006, xiv) notes, the "fundamentals of feminism have been recast in commercial terms. . . . The feminist ethic of economic independence has become the golden apple of buying power—a 'power' that for most women yields little more than credit-card debt, an overstocked closet, and a hunger that never gets sated."

In order to get technologies that are feminist, we must move from a "technology-push" approach to a "needs-led" one (Cockburn and Ormrod 1993). This would represent what I call a "cultural fix" (2000). At present, consumer "culture has tended to delight in clever new inventions and devices," in innovation for innovation's sake (Cockburn and Ormrod 1993, 80). Promoting "'only those [products] that have *benefits for people*'" would be a radical innovation (Cockburn and Ormrod 1993, 82, emphasis in original); promoting *only those that benefit women* would be more radical still.

One way to achieve this would be through Feminist Technology Assessments (FTA). Such assessments would be a development and extension of existing assessment tools including technology assessments, environmental impact statements, social impact statements, health impact assessments. A Feminist Technology Impact Assessment would be "a combination of procedures, methods and tools by which a policy, program or project may be judged as to its potential effects on [women], and the distribution of those effects [among women]" (adapted from European Centre for Health Policy 1999, 4). Such feminist assessments should become as common as these other types of assessment. In order to bring this into being, women need "to have key skills and the key jobs" so that we can "interrogate technological developments, assess them against various women's needs, and influence their course" (Cockburn and Ormrod 1993, 175).

Hardon (this volume) provides three models for feminist intervention into technological developments. In the case of Norplant, the intervention took place after a problematic new technology had already been developed and was being tried and promoted around the world; in the case of antifertility vaccines, the feminist intervention took place earlier in the process, while different versions of the technology were still being developed and tested; in the third case, that of microbicides, Hardon and other feminist health advocates were able to bring diverse groups of future female users into the design process early on so that they could be involved in setting the parameters for the technology during its development. The lesson is feminists should intervene early and often. Although intervening as early as possible is clearly preferable, Hardon demonstrates that it is not too late to do so even after a product has reached the market.

In addition to the three modes of engagement Hardon describes, there are other things we can do. One of these is to train feminist designers. Bronet and Layne (this volume) describe the techniques they have used and success they have had in teaching interdisciplinary design teams to design feminist products. Innovative courses like the feminist "Inventor's Studio" technology that historian Sharra Vostral and professor of industrial design Deana McDonagh (this volume) are instituting at the University of Illinois, which "revolve around designing and redesigning everyday objects to better serve women's needs," will surely help.

Another way we could encourage the design of feminist technologies would be to establish a foundation that would evaluate, promote, and publicize successful feminist technological innovations.[39] Such a foundation would award annual prizes for the best feminist technologies in a variety of areas or subfields, publicize the winners, and provide small grants to potential developers/designers.

In other words, there are many things we can do to promote feminist technologies. You do not have to be a designer or engineer to participate, although it certainly would be a great boon to have more feminist designers and engineers. Every one of us can use our buying power more strategically to encourage research and development of goods that would help us actualize our potential and help us lead more satisfying lives. Those of us who work in the media can publicize successes. Those who work in the nonprofit or public sectors can create programs and public policies that promote feminist technosocial systems. Scholars can help here too, for example, as Boyer and Penc (this volume) do when they identify and call for legislation that would protect women's right to have the time and suitable space provided for nursing or pumping breaks at work. Scholars and activists can participate in feminist technology assessment and can work with companies (like Medela, as Boyer [2007] proposes to do) or organizations (like the World Health Organization or Population Council, as Hardon has done) to make sure future women users are involved at all stages of the design process. There are plenty of opportunities for action and the time is ripe. Let's work together using our many talents to foster feminist technologies and technosocial systems that will eliminate gender-based inequities and truly enhance women's lives.

NOTES

1. Despite their utter familiarity today, these terms have not been in use very long; neither "technology" nor "feminist" were commonplace before the last quarter of the twentieth century. Oldenzeil (1999, 186) describes the 1978 *Encyclopedia Britannica* entry on technology as a landmark definition and notes that its emphasis on "mechanical, civil, and mining engineering as the pivot of technological change" made it difficult to "perceive women in the technological project."

2. It was in this same essay that Bush introduced the idea that technologies have a "valence," like atoms that have lost or gained electrons through ionization, and "have tendencies to pull or push behavior in definable ways. . . . Guns, for example, are valenced to violence; the presence of a gun in a given situation raises the level of violence by its presence alone." Guns are "a technology that is designed for killing in a way that ice picks, hammers, even knives—, all tools that have on occasion been used as weapons,—are not" (Bush 1983, 154–55).

3. We revisited landmark volumes like Rothschild's 1983 *Machina Ex Dea: Feminist Perspectives on Technology*, Cowan's *More Work for Mother* (1983), Zimmerman's *The Technological Woman* (1983), Wajcman's *Feminism Confronts Technology* (1991), Cockburn and Ormrod's *Gender and Technology in the Making* (1993), Grint and Gill's *The Gender Technology Relation* (1995), and consulted more recent work including Terry and Calvert's *Processed Lives: Gender and Technology in Everyday Life* (1997), Hopkins's *Sex/Machine: Readings in Culture, Gender, and Technology* (1998), Oldenziel's *Making Technology Masculine* (1999), Rothschild's *Design and Feminism: Re-Visioning Spaces, Places, and Everyday Things* (1999), Saetnan, Oudshoorn, and Kirejczyk's *Bodies of Technology: Women's Involvement with Reproductive Medicine* (2000), Gorenstein's *Gender and Work* (2000), Wosk's *Women and the Machine* (2001), Lerman, Oldenziel, and Mohun's *Gender and Technology* (2003), Wajcman's *TechnoFeminism* (2004), and Fox, Johnson, and Rosser's *Women, Gender, and Technology* (2006).

4. For example, Rothschild's 1983 *Machina Ex Dea: Feminist Perspectives on Technology* (the collection in which Bush used the term "feminist technology") and her 1988 book *Teaching Technology from a Feminist Perspective*. Pursell's (2001) historical review of the effects of feminism in the history of technology describes several other early works that use feminism in their title, including Hayden's 1981 *The Grand Domestic Revolution: A History of Feminist Designs for American Homes, Neighborhoods, and Cities*, Smith's 1983 essay "Women and Appropriate Technology: A Feminist Assessment," and McGaw's 1989 essay "No Passive Victims, No Separate Spheres: A Feminist Perspective on Technology's History."

5. Notable exceptions are Rothschild's 1999 volume and Wajcman's two books (1991, 2004). Wajcman's aim in *Feminism Confronts Technology* (1991) was to "make a strong case for building a feminist perspective into social science debates about technology . . . [and] to show that artifacts are themselves shaped by gender relations, meanings and identities" (2004, vi). *TechnoFeminism* (2004) focuses on "the continuities and discontinuities between current an earlier feminist reflections on science and technology" and explores "the complex ways in which women's everyday lives and technological change

interrelate in the age of digitalization" (2004, vii, 6). Wajcman notes that "a recognition that gender and technoscience are mutually constitutive opens up fresh possibilities for feminist scholarship and action" but concentrates on scholarship, not action (2004, 8).

6. Not only does Creager, Lunbeck, and Schiebinger's volume begin with the word "FEMINISM" (capital letters in original), it is in a font size much larger than the rest of the title and it occupies the entire first line.

7. I did so in response to Lynn Morgan's invitation to participate in the panel on hormones she and Elizabeth Roberts organized for the annual meeting of the Society for Social Studies of Science in Paris.

8. "Technology," according to Volti (1988, 4), was first coined by Harvard professor Jacob Bigelow in the late 1820s. It is derived from the "ancient Greek word 'techne,'" which is translated as "'art,' 'craft,' or 'skill,'" which in turn is derived "from an even more ancient Indo-European root, tkes-, which meant to weave or fabricate." For other definitions, see Winner (1977, 4–12).

9. Layne, Vostral, and Boyer organized a double panel at the 2005 Social Studies of Science Society meeting in Pasadena and Layne organized a panel at the 2005 Society for Literature, Science, and Art meeting in Chicago. In addition to those included in this volume, our panels included a paper by Kyra Landzelius about advertisements for neonatal intensive care equipment, a paper by Julie Wosk about new electronics being marketed to women, and a paper by Carol Collatrella about representations of women and technology in science fiction. The paper Bronet presented was about her work on a design course she taught on mobile incubators. Anita Hardon, Judy Wajcman, and Shirley Gorenstein served as commentators.

10. The volume is also focused, geographically and historically, on twentieth- and twenty-first-century North America. We are aware of how crucial historical/geographical/cultural differences are. By holding somewhat constant these variables, it allows us to develop our analysis more thoroughly. Rather than reifying or essentializing women, this focus, in fact, enables us to focus on less obvious forms of difference among women. On the one hand, it is possible that the very same women might use Seasonale, tampons, home pregnancy tests, and breast pumps. At the same time, these case studies illuminate important differences between contemporary North American women. The chapter by Hardon, while holding steady the focus on "feminine technologies," moves the geographical scope outward to ways of intervening in the international women's health movement. Together these studies provide an excellent launching pad for promoting feminist technology design.

11. We also discovered some technological innovations that may harm women because they only have the appearance of having solved a problem. An example of this is the blue light telephones, a public safety system installed on many college campuses that include a traceable panic button and purportedly create a safe public environment for female students. As Dusyk (2002), a master's student in RPI's STS Department observed, these technologies (which did not exist in her home country of Canada), are ill-conceived because the biggest threat to women's safety on college campuses is acquaintance violence, which takes place indoors in private or semiprivate spaces. Blue light telephones are not

really safety technologies but PR devices directed at prospective students and parents. They are primarily communication devices; not to communicate that a woman is being attacked and needs help, but rather "physical manifestations of 'something being done about violence.'"

12. Gorenstein credits Johnson with this insight. Gorenstein adds, "If we see the gussied-up cell phone as an extension of women's bodies, then we can say it is rooted in an attempt to increase the sexual dimorphism (difference in form of sexes in same species) of men and women"—with the implication being that women have tended to fare better when sexual dimorphism is downplayed rather than accentuated.

13. Another example of feminization of consumer products aimed at women is documented by Faludi. She describes how in 1987 "high femininity" or "frou-frou" "little girl dresses" were being pushed on women at a time when the average American woman was thirty-two years old. She attributes this to "an eruption of long-simmering frustration and resentment at the increasingly independent habits of modern female shoppers who "don't do as they're told anymore." A large study of women's fashion-shopping habits in the early '80s found that the more confident and independent women become, the less they liked to shop; the more they enjoyed their work, the less they cared about their clothes. The 1987 fashion also was a response to the way that annual sales of women's suits had been rising while dresses declined, which resulted in a net loss for the fashion industry because "business suits weren't subject to wild swings in fashion and women could get away (as men always have) with wearing the same suit for several days and just varying the blouse and accessories" (2006, 188). In an effort to force women to buy more dresses, the U.S. apparel manufacturers cut their annual production of women's suits.

14. Clearly the flood of pink items has been fueled in part by the breast cancer awareness movement. Ehrenreich (2001) vividly describes the pink kitsch that permeates the movement.

15. Samsung launched another phone designed specifically for women in 2003, the "T500 mobile with cherry red lacquer and light-up flashing crystals." More than 20,000 were sold in Australia (Manktelow 2005).

16. Another illustration of the way women's toilet needs can be a hardship for women on the job can be found in the class-action sexual harassment case of *Jenson v. Eveleth Taconite Co.*, against an iron mine in northern Minnesota depicted in the Warner Brothers movie *North Country*. In response to a presentation of this introduction, a member of the audience at Liverpool University recommended a "she pee," a device she had purchased at a walking store and had found very useful while taking hikes. This technology had also been suggested by one of the PDI students described in Bronet and Layne (this volume). But must women urinate like men to enjoy the freedom men do? (See chapter 7 for a discussion of the social acceptance of public displays of bodily fluids depending on gender.)

17. The rate of nonpromotion for men declined from 21 to 16 percent.

18. Perhaps such a technology qualifies for the least feminist on a continuum of technologies that are feminist, as is discussed later in this essay, especially if those women who are aided by it then use their new power/authority to help other women.

19. This helps explain why other feminists oppose or are skeptical of contraceptives that require women to trust men with their fertility (Oudshoorn 2000, 136).

20. "Of the 299 Pitt County defendants arrested for domestic violence in 2004 . . . 36 percent were involved in one or more additional domestic violence incidents before they were brought to trial and 20 percent were rearrested. In the worst case, deputies responded 19 times to domestic violence incidents involving one man before his trial on the first charge. On nine of those calls, he was arrested" (Weigl 2006).

21. One patent (U.S. Patent 5396227) for a portable electronic monitoring system warns "the victim and authorities, and automatically gathers evidence, independent of any that may be provided by the victim of the violation." In other words, one of the rationales for this innovation is that the testimony of the woman/victim is not considered authoritative.

22. One feminist has proposed a version of this technology that would be used by women (instead of the male perpetrator). She suggests adapting medical alert systems like the necklaces worn by the elderly combined with a GPS locating system. A woman's "cry for help would be routed to her local rescue volunteers . . . like a volunteer fire department. There could be brigades in every city and town with the very same commitment to saving human life that fires call upon" (Rich 1997, 230). Rather than relying solely on teaching women self-defense (which places the responsibility on each woman individually), she prefers the rescue model, in which members of a community take responsibility for each other's safety.

23. Research on television remote control use shows that men use the devices more frequently than women, are more likely to use them for grazing and multiple channel viewing, and in mix gender viewing situations, men are more likely to control the device (Bellamy and Walker 1996:142).

24. Bell's company also provides computer-assisted design (CAD) services for the apparel industry.

25. The Web sites for these products are www.pumpease.com, www.babybuddha.ca, www.YOUpillows.com, www.prenatalcradle.com, www.lunapads.com, www.birthinbinsi .com, and www.discoverbirth.com. Discover Birth was started by Stefanie Antunes, who worked in business in the area of "competitive intelligence—helping companies understand their business environment, their competitors, and how the impending trends would impact them." After a disappointing birth, she became a certified Lamaze instructor and a doula, then she joined with other childbirth educators in Ontario (Canada) to form this company. Their slogan is "Changing the world . . . one family at a time" and their mission is to help other like-minded birth-related entrepreneurs succeed (www.discoverbirth.com). The *Wall Street Journal* and the SBA estimated in 1999 that 65 percent to 70 percent of new businesses will fail within the first five to eight years of operations. (http://askville.amazon .com/percentage-start-companies-years/AnswerViewer.do?requestId=9620767).

26. A Dutch journalist commenting on the large-scale clinical trial of contraceptive hormonal injections for men found it inconceivable that a woman would put herself in a situation in which she would need to either trust or try to "control [her] partner: 'Darling, did you take your injection?'" (quoted in Oudshoorn 2000, 136).

27. Although Johnson (this volume) does not arrange them along a continuum, her four

candidates for feminist technology (are good for women; constitute social relations that are more equitable than those that were constituted by a prior technology or those that prevail in the wider society; constitute gender-equitable social relations; or favor women) could be so arranged and used.

28. Personal communication, Frances Bronet, August 8, 2008.

29. The "traditional cry room" is a closed room, typically located at the back of a sanctuary. Some versions offer a full day care located along the side of the sanctuary separated by a glass wall and equipped so the children hear "the worship through a sound system" and parents can "keep an eye on their children while also participating in the service" (Bott 2007).

30. To attenuate the isolation problem, a new design has been introduced, that of the "walk-in cry room," which does not have a door but whose entrance resembles that of many public restrooms. As a result, "a loud infant will still be heard . . . but only as a distant, muted presence that is not distracting to most people . . . this design connects the two areas, acoustically [and] . . . more importantly . . . emotionally, reducing feelings of isolation (Bott 2007).

31. According to the docent, young children were seated with their backs to the pulpit, facing adult family members; teenagers sat in the balcony—girls on one side, boys on the other.

32. As a write-up of such a club in an online parenting magazine states, this "is a perfect way for having a couple of hours in an adult environment and allows you to keep up with the latest blockbusters without having to pay for a babysitter or worrying about your baby causing a disturbance" (Allen 2006). Rather than an expensive single-use structural alteration in the space (like a cry room), the same spaces are tweaked for a new use. "This enterprising scheme" (Allen 2006) clearly has the potential for increasing revenues by finding a new market for the facility during off-use hours.

33. The benefits of politically informed product assessment and redesign can be seen in the case of diapers. For years the environmental pros and cons of reusable cloth versus disposable diapers have been publicly debated (Berk 1997; New Parents Guide 2008; Onion 2005). Since the 1970s and '80s, when throw-away diapers became the norm, opponents have objected to the environmental damage they do. In response, disposable diaper manufacturers and users have pointed out the environmental costs of raising, manufacturing, and transporting cotton and the resources used to launder them. The result of the debate has been that both have been improved: disposables have been developed that are more easily biodegradable and may use fewer or less-harmful chemicals to produce them while reusable cloth diapers have been made more convenient through innovative designs and greener through the advent of safer detergents and energy-saving washers and dryers (example suggested by Bronet, personal communication 2008).

34. See McCaughey (in press) for a reflexive account of a science and technology studies scholar's experience of breast-feeding in the midst of these feminist tensions.

35. A series of three programs on "Girls and Their Gadgets" aired on National Public Radio in June 2008 (2008a, 2008b, 2008c).

36. Families that do practice "equally shared parenting" often find that using the simple

technology, pioneered by the founders of ThirdPath, of a color-coded computer chart to track who is home during the week helps (Belkin 2008, 48).

37. In contrast, "American female investors tend to avoid risk" while "their Japanese counterparts are taking enormously risky, speculative trades" (Schultz 2008, 5).

38. The article fails to mention the high rate of poverty in single-mother families.

39. The proposed GEM Foundation described in chapter 7 by Bronet and Layne (this volume) could be adapted to provide a template.

References

Allen, Mary. 2006. London for Free; Families in South West London. www.familiesonline .co.uk. (Accessed March 25, 2006.)

Baby Buddha. www.babybuddha.ca.

Belkin, Lisa. 2008. "When Mom and Dad Share It All." *New York Times Sunday Magazine.* June 15, pp. 44–51.

Bellamy, Robert V. Jr., and James R. Walker. 1996. *Television and the Remote Control: Grazing on a Vast Wasteland.* New York: Guilford Press.

Berk, Luanne. 1997. The Ecological Debate: Cloth vs. Disposable. www.geocities.com/ Wellseley/Atrium/8608/cloth-vs-disposable.html. (Accessed March 25, 2008.)

Berkley, Robyn, Cheryl Geisler, Linda Layne, and Deborah Kaminski. N.d. The Effects of Gender on the Promotion to Full Professor: A Case Study of a Technological University. Unpublished manuscript.

Bott, Tim. 2007. Cry Room Acoustics: How to Make Your Cry Room Both Comfortable and Acoustically Sound. Church Acoustics/Acoustic Sciences Corp. www.church-acoustics .com/cryroom.htm. (Accessed April 2, 2008.)

Boyer, Laura Kate. 2007. "The Role of User Feedback in Breast Pump Design." Grant application to the British Academy.

Burgess, Gemma. 2008. *Gender Equality Duty in Urban Regeneration Relay: The Newsletter of Research in the School of the Humanities and Social Sciences* 4 (May). University of Cambridge.

Bush, Corlann Gee. 1983. "Women and the Assessment of Technology: To Think, to Be, to Unthink, to Free." In Joan Rothschild, ed., *Machina Ex Dea: Feminist Perspectives on Technology,* 151–70. New York: Pergamon.

Campbell, Kim. 2003. "Different Rulers—Different Rules." In Deborah L. Rhodes, ed., *The Difference "Difference" Makes: Women and Leadership,* 121–28. Stanford, Calif.: Stanford University Press.

Caporael, Linnda R., Panichkul, E. G., and Harris, D. R. 1993. "Tinkering with Gender." Special issue, Joan Rothschild, ed., "Technology and Feminism," *Research in Philosophy and Technology* 13: 73–99. Greenwich, Conn.: JAI Press.

Cleveland, Amber. 2007/2008. "Innovation Education: Changing the Way Design Impacts the World." *Rensselaer Alumni Magazine* (Winter): 24–29.

———. 2008. Rensselaer Students Recognized for Innovative Ideas to Change the World.

Press release. http://news.rpi.edu/update.do?artcenterkey=2424. (Accessed April 16, 2008).

Cockburn, Cynthia, and Susan Ormrod. 1993. *Gender and Technology in the Making.* London: Sage.

Cowan, Ruth Schwartz. 1983. *More Work for Mother. The Ironies of Household Technology from the Open Hearth to the Microwave.* New York: Basic Books.

Creager, Angela N. H., Elizabeth Lunbeck, and Londa Schiebinger. 2001. *Feminism in Twentieth-Century Science, Technology, and Medicine.* Chicago: University of Chicago Press.

Despois, Emilie. 2005. Baby Scream Families in South West London. www.familiesonline .co.uk (Accessed April 5, 2005).

Discover Birth. www.discoverbirth.com.

Dusyk, Nicole. 2002. Safety Designs and Gendered Constructions: Framing Violence against Women through Technology and Policy. Unpublished Internship Report for the MS in STS, RPI.

Ehrenreich, Barbara. 2001. "Welcome to Cancerland: A Mammogram Leads to a Cult of Pink Kitsch." *Harpers Magazine* (November): 43–53.

Equality and Human Rights Commission. 2008. Gender Equality Duty. www .equalityhumanrights.com/en/forbusinessesandorganisation/publicauthorities/ Pages/PublicauthoritiesGenderequalityduty.aspx. (Accessed May 22, 2009.)

European Centre for Health Policy. 1999. *Health Impact Assessment: Main Concepts and Accepted Approach.* Gothenburg Consensus Paper. Brussels.

Everts, Saskia. 1998. *Gender and Technology: Empowering Women, Engendering Development.* New York: St. Martin's Press.

Faircloth, Charlotte. 2009. Mothering as Identity-Work: "Full-Term" Breastfeeding, Attachment Parenting and Intensive Motherhood. Ph.D. thesis, Department of Social Anthropology, University of Cambridge.

Faludi, Susan. 2006. *Backlash: The Undeclared War against American Women* (15th anniversary ed.) New York: Three Rivers Press.

Fox, Mary Frank, Deborah G. Johnson, and Sue Rosser, eds. 2006. *Women, Gender, and Technology.* Champaign: University of Illinois Press.

Geisler, Cheryl, Debbie Kaminski, and Robyn A. Berkley. 2007. "The 13+ Club: A Metric for Understanding, Documenting, and Resisting Patterns of Non-Promotion to Full Professor." *National Women's Studies Association Journal* 19, no. 3: 145–62.

Gogoi, Pallavi. 2005. "'I Am Woman, Here Me Shop': Rising Female Consumer Power Is Changing the Way Companies Design, Make, and Market Products—and It's about More Than Adding Pastels." *Business Week* (February 14). http://www.businessweek .com/bwdaily/dnflash/feb2005/nf20050214_9413_db_082.htm. (Accessed September 2, 2008.)

Gorenstein, Shirley. 1996. *Research in Science and Technology Studies: Material Culture.* Knowledge and Society Series. Stamford, Conn.: JAI Press.

———. 2000. *Research in Science and Technology Studies: Gender and Work.* Knowledge and Society Series. Stamford, Conn.: JAI Press.

———. 2005. Comments on panel "Feminist Technologies?" Society for Science, Literature and Art's annual meeting, Chicago.

———. 2007. Comments on first draft of the manuscript *Feminist Technology.*

Grint, Keith, and Rosalind Gill. 1995. *The Gender-Technology Relation: Contemporary Theory and Research.* London: Taylor and Francis.

Hayden, Dolores. 1981. *The Grand Domestic Revolution: A History of Feminist Designs for American Homes, Neighborhoods, and Cities.* Cambridge, Mass.: MIT Press.

Herszenhorn, David M. 1999. "Alarm Helps to Fight Domestic Violence." *New York Times* (July 27).

Hopkins, Patrick D., ed. 1998. *Sex/Machine: Readings in Culture, Gender, and Technology.* Bloomington: Indiana University Press.

Johnson, Deborah. 2006. Introduction. In Mary Frank Fox, Deborah G. Johnson, and Sue Rosser, eds., *Women, Gender, and Technology,* 1–5. Urbana: University of Illinois Press.

Krotz, Joanna L. 2008. Women Power: How to Market to 51 Percent of Americans. Microsoft Small Business Center. http://www.microsoft.com/smallbusiness/resources/marketing/market-research/women-power-how-to-market-to-51-of-americans.aspx#WomenpowerhowtomarkettoofAmericans. (Accessed September 5, 2008.)

Layne, Linda L. 2000. "'The Cultural Fix': An Anthropological Contribution to Science and Technology Studies." *Science, Technology, and Human Values* 25, no. 4: 492–519.

———. 2003. *Motherhood Lost: A Feminist Account of Pregnancy Loss in America.* New York: Routledge.

———, and Heather Bailey, co-producers. 2006. "Preparing for Home Pregnancy Loss: A Conversation with Sandy Maclean of WomenCare," an episode of *Motherhood Lost: Conversations.* Fairfax, Va.: George Mason University–TV.

Lerman, Nina E., Ruth Oldenziel, and Arwen P. Mohun, eds. 2003. *Gender and Technology: A Reader.* Baltimore: Johns Hopkins University Press.

Lewin, Ellen. 1993. *Lesbian Mothers: Accounts of Gender in American Culture.* Ithaca: Cornell University Press.

Maines, Rachel. 1999. *The Technology of Orgasm: "Hysteria," Vibrators and Women's Sexual Satisfaction.* Baltimore: Johns Hopkins University Press.

Manktelow, Nicole. 2005. What Women Want. Livewire-Technology. http://www.theage.com.au/news/livewire/what-women-want/2005/11/30/1133026497218.html#. (Accessed May 20, 2009.)

Martinelli, Nicole. 2007. "What Do Women Want? Less Pink, More Tech." *Wired* (September 10). http://www.wired.com.

McCaughey, Martha. In press. "Got Milk?: A Feminist STS Scholar's Experience of Breast Feeding." Special issue, Menjamin Cohen and Wyatt Galusky, eds., *Science as Culture on Living Narratives.*

McGaw, Judith A. 1989. "No Passive Victims, No Separate Spheres: A Feminist Perspective on Technology's History." In Stephen H. Cutcliffe and Robert C. Post, eds., *Context: History and the History of Technology,* 172–91. Bethlehem, Pa.: Lehigh University Press.

———. 2003/1996. "Why Feminine Technologies Matter." In Nina E. Lerman, Ruth Oldenziel, and Arwen P. Mohun, eds., *Gender and Technology: A Reader*, 13–36. Baltimore: Johns Hopkins University Press.

Melendez, Michele M. 2006. "Companies Size Up Potential in Items Specific to Women: Fit with Tools, Sports Gear Key in Performance, Safety." *San Diego Union-Tribune* (July 2). Newhouse News Service. www.signonsandiego.com/uniontrib/20060702/news_1002gean. html. (Accessed September 20, 2008.)

Meyer, Donald. 1987. *Sex and Power: The Rise of Women in America, Russia, Sweden, and Italy.* Middletown, Conn.: Wesleyan University Press.

Mohr, Hans. 1999. "Technology Assessment in Theory and Practice." *Philosophy and Technology* 4, no. 4: 22–25.

National Public Radio. 2008a. The Most Ridiculous Girl Gadgets. *Day to Day* (June 9). www.npr.org.

———. 2008b. What Women Want in a Cell Phone. *Day to Day* (June 10). www.npr.org.

———. 2008c. Why the Audrey Gadget Flopped. *Day to Day* (June 11). www.npr.org.

New Parents Guide. 2008. Diapers, Diapers and More Diapers: Cloth vs. Disposable (June 10). www.thenewparentsguide.com/diapers.htm.

Oldenziel, Ruth. 1999. *Making Technology Masculine: Men, Women and Modern Machines in America 1870–1945.* Amsterdam: Amsterdam University Press.

Onion, Amanda. 2005. "The Diaper Debate: Are Disposables as Green as Cloth?: New British Study Adds to Conflicting Conclusions on the Greenest Way to Diaper Your Baby's Bottom. *ABC News* (May 26). www.abcnews.go.com/Technology/Story?id=789465&page=1. (Accessed Sept. 8, 2008.)

Oudshoorn, Nelly. 2000. "Imagined Men: Representations of Masculinities in Discourse on Male Contraceptive Technology." In Ann Rudinow Saetnan, Nelly Oudshoorn, and Marta Kirejczyk, eds., *Bodies of Technology: Women's Involvement with Reproductive Medicine*, 123–45. Columbus: Ohio State University.

———, Ann R. Saetnan, and Merete Lie. 2002. "On Gender and Things: Reflections on an Exhibition on Gendered Artifacts." *Women's Studies International Forum* 25, no. 4: 471–83.

———. 2003. *The Male Pill: A Biography of a Technology in the Making.* Durham, N.C.: Duke University Press.

———, and Trevor Pinch, eds. 2003. *How Users Matter: The Co-Construction of Users and Technology.* Cambridge, Mass.: MIT Press.

Our Hope Place. www.ourhopeplace.com.

PumpEase, Hands-Free Pumping Support. www.pumpease.com.

Pursell, Carroll. 2001. "Feminism and the Rethinking of the History of Technology." In Angela N. H. Creager, Elizabeth Lunbeck, and Londa Schibinger, eds., *Feminism in Twentieth-Century Science, Technology, and Medicine*, 113–27. Chicago: University of Chicago Press.

Rensselaer Polytechnic Institute. 2007. Women at Rensselaer. www.lib.rpi.edu/Archives/ gallery/women. (Accessed September 3, 2008.)

———. 2008a. Rensselaer's History. http://www.rpi.edu/about/history.html. (Accessed September 3, 2008.)

———. 2008b. E*ntrepreneurship at Rensselaer. http://www.rpi.edu/about/entrepreneurship .html. (Accessed May 28, 2008.)

Rich, B. Ruby. 1997. "The Party Line: Gender and Technology in the Home." In Jennifer Terry and Melodie Calvert, eds., *Processed Lives: Gender and Technology in Everyday Life*, 222–31. New York: Routledge.

Rothschild, Joan. 1983. *Machina Ex Dea: Feminist Perspectives on Technology*. New York: Pergamon.

———. 1988. *Teaching Technology from a Feminist Perspective: A Practical Guide*. Elmsford, N.Y.: Pergamon.

———. 1989. "From Sex to Gender in the History of Technology." In Stephen H. Cutcliffe and Robert C. Post, eds., *Context: History and the History of Technology*, 191–203. Bethlehem, Pa.: Lehigh University Press.

———. 1999. *Design and Feminism: Re-Visioning Spaces, Places, and Everyday Things*. New Brunswick, N.J.: Rutgers University Press.

Saetnan, Ann Rudinow. 1999. The Gender of Things II—Public Response in Trondheim. http://www.svt.ntnu.no/iss/Ann.R.Saetnan/gender.htm. (Accessed Sept 2, 2008.)

———, Nelly Oudshoorn, and Marta Kirejczyk, eds. 2000. *Bodies of Technology: Women's Involvement with Reproductive Medicine*. Columbus: Ohio State University.

Sanabria, Emilia. 2008. Limits That Do Not Foreclose: Biomedical Intervention, Hygiene and Sex Hormones in Salvador, Brazil. Ph.D. dissertation in Social Anthropology, King's College, Cambridge.

Schroeder, Patricia Scott. 2003. "Women's Leadership: Perspectives from a Recovering Politician." In Deborah L. Rhode, ed., *The Difference "Difference" Makes: Women and Leadership*, 85–89. Stanford, Calif.: Stanford University Press.

Schultz, Emily. 2008. Feminism and Women's Use of On-Line Investment Technologies in Japan and the United States. Unpublished senior thesis, Department of STS, RPI.

Sexual Harassment Support, a Support Community for Anyone Who Has Experienced Sexual Harassment. www.sexualharassmentsupport.org.

Sklar, Kathryn Kish. 1998. "Catherine Beecher (1800–1878)." In G. J. Barker-Benfield and Catherine Clinton, eds., *Portraits of American Women from Settlement to the Present*, 169–88. New York: Oxford University Press.

Smith, Judy. 1983. "Women and Appropriate Technology: A Feminist Assessment." In Jan Zimmerman, ed., *The Technological Woman: Interfacing with Tomorrow*, 65–70. New York: Praeger.

Stimpson, Catharine R. 2001. Foreword. In Angela N. H. Creager, Elizabeth Lunbeck, and Londa Schibinger, eds., *Feminism in Twentieth-Century Science, Technology, and Medicine*, vii–x. Chicago: University of Chicago Press.

S. W. 2008. "Tased and Confused." *Marie Claire* (April): 116.

Terry, Jennifer, and Melodie Calvert. 1997. *Processed Lives: Gender and Technology in Everyday Life*. New York: Routledge.

Tujak, Laura. 2008. "A Book Review of *Knock Yourself Up* by Louis Sloan." *Single Mothers by Choice* 103:8.

Vanclay, Frank. 2003. *Social Impact Assessment: International Principles International Association for Impact Assessment.* Special Publication Series 2. Fargo, N.D.

Volti, Rudi. 1988. *Society and Technological Change.* New York: St. Martin's Press.

Vostral, Sharra L. 2008. *Under Wraps: A History of Menstrual Hygiene Technology.* Lanham, Md.. Rowman & Littlefield.

Wajcman, Judy. 1991. *Feminism Confronts Technology.* University Park, Pa.: Pennsylvania State University Press.

———. 2004. *TechnoFeminism.* Cambridge: Polity.

Weigl, Andrea. 2006. GPS Anklets Track Battering Suspects. Blueline Radio (January 1). http://www.bluelineradio.com/GPS.html.

Winner, Langdon. 1977. *Autonomous Technology: Technics Out of Control as a Theme in Political Thought.* Cambridge, Mass.: MIT Press.

———. 1986. *The Whale and the Reactor: A Search for Limits in an Age of High Technology.* Chicago: University of Chicago Press.

———. 2002. Gender Politics and Technological Design. Paper presented at the Women's Studies Center, Colgate University.

Woodhouse, Edward. 2008. Woodhouse's Law. Unpublished course handout.

Wosk, Julie. 2001. *Women and the Machine: Representations from the Spinning Wheel to the Electronic Age.* Baltimore: Johns Hopkins University Press.

Young, Linda. 2008. "Study: Marriage Creates Seven Hours Extra Housework for Women per Week." AHN Media Corp. www.allheadlinenews.com/articles/. (Accessed December 12, 2008.)

Zimmerman, Jan, ed. 1983. *The Technological Woman: Interfacing with Tomorrow.* New York: Praeger.

1   Sorting Out the Question of Feminist Technology

DEBORAH G. JOHNSON

IN THE PAST TWO DECADES, scholarship on the relationship between gender and technology has grown significantly. Drawing on concepts and theories in technology studies, a major focus of the literature on gender and technology has been directed at understanding how a system of gender relations becomes inscribed in a technology and, vice versa, how technology reinforces, embodies, or disrupts gender ideas and relationships. Lurking in the background of much of the literature on gender and technology is an interest in social change, social change that will improve the circumstances of women and create more equitable gender relations. Indeed, one might say that lurking in the background is a question about whether technology is a friend or foe of feminism. This is not to say that those who study technology and gender are always suspicious of, or negative about, technology. There is often an implicit suggestion of just the opposite, that if we better understood the relationship between gender and technology, we might use that understanding to further the progress of feminist social change.

In this chapter, I want to bring these background ideas and possibilities to the fore, and confront the connection not just between gender and technology, but also between feminism and technology. I do this by pursuing the question of this volume, the question whether there are or could be feminist technologies. The question is, to be sure, somewhat odd—what could it mean to say that a technology is feminist? Would it mean, for example, that the technology can only be used for feminist goals? Would a feminist technology only lead to positive consequences for women? Or what? The question may also be somewhat dangerous because it seems to call for an essentialist answer. An answer could fall into the trap of essentialism both about women and about feminism. Would a feminist technology be one that is good for all women? Would it be one that realized the goals and interpretations of all feminisms? To be sure, these dangers should be avoided. The analysis provided here tries to continually question the meaning of the question while at the same trying to answer it. The point of asking the question is to see where it takes us. The analysis is exploratory: it is a first attempt by this philosopher to sort out a concept and frame it in interdisciplinary, theoretical territory. Exploring the question will, hopefully, shed new light on the feminism–technology connection.

ASSOCIATIONS OF GENDER AND TECHNOLOGY

As much of the scholarship on gender and technology indicates, the starting place for thinking about the intertwining is with the cultural associations between gender and technology. These provide an important backdrop for approaching the feminist technology question. The association relationship is far from simple. But in Euro-American culture there are several important and conflicting associations worth mentioning. First, technology is often associated with masculinity. Technology is thought to be masculine—the domain of the male—while women are often thought to be inept with technology, ignorant and unskilled with regard to how artifacts work, and simply less interested in it. Woman—the feminine—is associated with nature, and technology is just the opposite; it is the realm of the human-made, the artificial. Despite social programs and policies aimed at increasing the number of women who enter the field of engineering—the domain of expertise with regard to technology—engineering is still associated with men. The gender composition of the field continues to support this association in that there are still relatively small numbers of women pursuing careers in engineering.

Part of the association between technology and masculinity is, of course, constructed through selective use of the term "technology" (Stanley 1998). Human-made, material objects used by men are called technology; human-made, material objects used by women are referred to as tools or utensils or appliances. Domains of knowledge and skill mastered by men are called technical or technological while those mastered by women are considered crafts. In this respect, while the cultural association of technology with masculinity is undeniable, it is important to remember that associations are cultural constructions with complex histories.

Current notions of what technology is can be traced back to the late nineteenth century when engineers, as part of the process of professionalization, were claiming special expertise. As engineers defined technology, the significance of both artifacts and forms of knowledge associated with women were left out (Wajcman 2004). Thus, the feminist technology question must take into account that although associations are real (e.g., people actually do associate technology with masculinity), the things associated can be understood and constructed in other ways.

The other type of association has to do with the way certain technologies are associated with women and others with men. This probably derives from the historical separation of spheres, setting up an association of certain domains of life with a particular sex. While these associations can and do change as gender systems change, certain tools or devices, skills, and domains of expertise are associated with men and others are associated with women. Men are expected to know, and are more likely to know, about how automobiles and electronic devices work and know how to fix "things"; women are expected to know more about child-rearing equipment and techniques, cooking techniques and tools, how to use cleaning appliances, and so on. These associations are well known from everyday life.

Of course, gender systems change, and so do these associations. Thus, the strength of the association of particular technologies with men or with women may be weaker today, at least in the United States, than twenty years ago. At the same time, the stereotypes seem to have an uncanny persistence. We continue to think of doctors as male even though women doctors outnumber men in certain fields of medicine. We continue to think of the kitchen as women's domain while the number of men who cook and are quite interested in cooking is on the rise, not to mention that the so-called great chefs of the world are predominately male.

Nothing that has been mentioned so far is new or startling; it simply provides the background for approaching the question of feminist technology. Although it might seem a small step to move from certain technologies being considered feminine to certain technologies being considered feminist, the step is fraught with complexity.

CO-CONSTRUCTION OF GENDER AND TECHNOLOGY

The past two decades of scholarship in the field of science and technology studies (STS) have provided the concepts and tools to move well beyond the associations of gender and technology. The field seems to have converged around a thesis that provides a framework for the feminist technology question. Although there are many contentious fine points, STS theory is centered on the idea that technology and society co-constitute one another (Bijker 1995; Mackenzie and Wajcman 1999). STS scholarship has been directed at theorizing, modeling, explaining, or exploring how a wide variety of social factors, social conditions, cultural notions, ideologies, and policies shape the development and endurance of technology. And vice versa, STS theory explores how artifacts, techniques, and systems of knowledge and expertise shape and constitute society; artifacts are intertwined in social practices, social institutions, and cultural notions. Technology and society are inseparable; we can't really disentangle them.

An important corollary of the STS co-construction thesis is a revision to the concept of technology. Technology is not merely material objects or artifacts. Material objects do not and cannot exist or have meaning or use independent of social endeavors, social processes, social practices, and social meaning. Thus, when it comes to studying technology, the focus of attention and the unit of analysis should not be on artifacts alone, but rather *sociotechnical systems*. Technology *is* the combination of artifacts together with social practices, social relationships and arrangements, social institutions, and systems of knowledge. The combination is variously referred to as *sociotechnical ensembles* (Bijker 1995) or *sociotechnical systems* (Hughes 1994) or *networks* (Law 1987). In actor-network theory (ANT), artifacts are treated as actants, equal to human actors in their potential influence on the system (Callon 1986). While technology is not just artifacts, neither is the social just social. Many social practices, relationships, institutions, and arrangements are partly constituted by artifacts. Whether the social endeavor is work, education, child rearing, or medical care, artifacts are involved. Artifacts shape these social practices; social activities or practices could not exist as they do without the artifacts. Even something as profoundly social as child rearing is constituted in part by artifacts. Child rearing—at least as we know it today— consists of a wide array of social practices of engagement between parents and children, and these are mediated and constituted by such things as baby bottles, playpens, car seats, inoculations, teething rings, child-rearing manuals, diapers, toys, etc.[1] The artifacts have shaped and been shaped by the social ideologies and practices of child rearing. Hence, child rearing is a sociotechnical system.

Bringing this notion of technology as sociotechnical systems to the question of feminist technology both helps and complicates the task. It brings us closer to an answer because social arrangements, social practices, and relationships are the "stuff" of feminism and the "stuff" of sociotechnical systems. Indeed, Cockburn and Dilic describe this connection as a key premise of their 1994 volume, *Bringing Technology Home*:

> The social constructivist approach shifts attention away from the artefact as hardware to the knowledge and processes that together give a thing meaning. All of us inevitably consolidated around this approach because of the demands made by the concept of "gender." Gender was clearly social relations. To be able to see gender and technology shaping one another, technology had to be seen as social relations too. (Cockburn and Dilic 1994, 7–8)

If we identify technology as social relations, a seemingly quick and easy answer to the feminist technology question appears: *feminist technologies are technologies that constitute and are constituted by feminist social relations.*

If only it were so simple! Unfortunately, this simple answer thrusts us into the middle of the most daunting questions of feminism and feminist theory. It also has the danger of pushing the materiality of technology completely out of sight.

MATERIALITY

Remember that sociotechnical systems consist of social relations together with artifacts. One of the most intriguing aspects of the feminist technology question is precisely that it calls upon us to consider the role of artifacts and the human-made material world in gender relations. This in turn raises the question whether the world has been physically constructed in ways that constrain feminist goals and it points to the possibility of design—feminist design—as a new strategy for feminist goals. Thus, although a focus on social relations allows us to consider technology and feminism together, we should not lose sight of artifacts and the built environment.

The role of artifacts is somewhat controversial within STS. Many STS theorists agree that artifacts are seen as artifacts or entities only through a lens of social and cultural meaning (Collins and Yearley 1992); they are delineated as "things" in social contexts that, in effect, make them things. Others hold on to the notion that the materiality of artifacts has some influence independent of how they are individuated and socially constructed (Callon and Latour 1992). Artifacts have particular (hard, nonmalleable) design features and whatever their cultural meaning, the design of the artifact can itself have an influence on social behavior.

Because answering the feminist technology question is an exploratory endeavor, I do not want to rule out either approach. It seems better to keep the status of artifacts alive and see what the feminist technology question reveals about it. This means, however, that we have to break the feminist technology question into two questions. One question has to do with artifacts—the materiality of technology and the design of artifacts. A second question has to do with sociotechnical systems, networks of social arrangements, practices, and relationships *together with* artifacts. Although pure constructivists may dismiss the first question outright, as I have already indicated, I find it intriguing precisely because the materiality of the human-made world is something that has not been fully addressed by feminism and seems a potentially important new site for feminist social action and social change. The second question, while important in its own right, is ultimately essential to understand the first question because artifacts are always components in sociotechnical systems. Whatever influence the materiality of the artifact has will operate in the context of the system of which it is a part. We have, then, two questions: (1) Are there (or could there ever be) artifacts that could be considered feminist? and (2) Are there (or could there ever be) sociotechnical systems that could be considered feminist?

THE DAUNTING QUESTIONS OF FEMINISM

When we claim that feminist technologies are those that constitute or are constituted by feminist social relations, we are thrust into the middle of the most daunting questions of feminism. What is feminism? Yes, feminism is for the empowerment of women but does this mean equality or does it mean more power for women than for men? Is feminism merely (or primarily) a critical perspective, identifying what is wrong (e.g., inequities) but not able to give content to the idea of alternative, nongendered social relations? Can we, who are situated here and now, even imagine a world in which feminist goals are achieved? What might that world look like? It might be equitable, but what would happen to sex and gender?

Compounding this problem is the enormous variation of women's circumstances. Feminists and feminist theory acknowledge that because of the varying circumstances of women as well as varying ideas about gender and strategies for improving the conditions of women, there may not be a single feminism but rather different forms of feminism (Frye 2000). This is to say that we must eschew essentialism with regard to feminism.

Yet the feminist technology question requires some sort of account of feminism, some characterization of the kind of social relations that could be considered feminist. If we cut to the core and try to characterize feminism in a way that spans different forms of feminism, several candidates seem ready at hand. First, we might say that feminism is about improving the conditions of women. Here feminist social relations would be those that are good for women. While there seems something right about this account, it doesn't seem to get at the gender component. That is, there are many social conditions and social relations that are good for women but have nothing to do with gender or may involve differential treatment between men and women.

A second candidate is gender equity. Feminism can be understood to be a view of the world that focuses on the inequities in women's circumstances as compared to men's and has as its goal creating a world in which no disadvantage correlates with being a woman. On this account, feminist social relations might prevail in a world in which social relations are generally cruel, selfish, and inhumane. But if women and men were alike treated in this way, feminism will have been realized. The point of focusing on equity is to emphasize that feminism is concerned with women's circumstances relative to men's or, perhaps more accurately, in relation to and with men. This is quite a contrast to feminism as concerned with what is good for women.

A third candidate arises when we imagine those who might challenge gender equity as much too weak an account of feminism. Perhaps our focus should be on technologies that favor women, rather than merely equalize. At the end of her piece on appropriate technology and women in the third world, Oblepias-Ramos suggests: "[I]t may be said that technologies become appropriate when they carry a deliberate bias for a specific underprivileged sector of a community, as well as an overall appreciation of that sector's overall physical and cultural environment" (1998, 93).

If we think of feminist technology as a kind of appropriate technology, we could think of it as technology that favors women. (Note here that it is not enough to say that feminist technology is technology that addresses women or women's needs because that is what is meant by "feminine" technology.) Technology that favors women could be understood, for example, to be technology that counters preexisting imbalances in gender relations, imbalances that favor men. Another reason for thinking of feminist technology this way is that many current technologies—artifacts, at least—seem to be gender neutral and, therefore, arguably, gender equitable, but calling them feminist does not seem right. Roads, clocks, furnaces, and telephones, for example, do not seem to address women and men differently, yet these artifacts do not seem good candidates for feminist technology. So, there is a point to thinking about feminist technologies as those that favor women in some way.

Another complexity comes into sight when we consider new technologies. Here we have to keep in mind that the new artifact or system will function in contexts in which a gender system existed before or, if the new artifact creates a new context, we have to remember that it functions in the context of a broader society in which a gender system operates. Because of this—that is, because gender systems prevail in the context in which technologies are newly introduced or in the broader society—the feminist technology question could be embraced simply as a relativistic notion. That is, a technology might be considered feminist when it improves upon prior conditions or creates social relations that are better for women than those in the broader culture. Whether the technology is feminist would, then, be a matter of whether the technology in question constituted more gender-equitable social relations than those constituted by a prior technology or more gender-equitable social relations than in the broader culture.

This relativistic account of feminist technology is very tempting; however, it seems to settle for too little. I said earlier that the point of probing the feminist technology question was to garner insights into the gender and technology connection in the hope of discovering more effective ways of intervening to improve the circumstances of women. In this respect, figuring out what could constitute feminist social relations is not something we should avoid.

We have then four candidates for feminist technology: (1) technologies that are good for women; (2) technologies that constitute gender-equitable social relations; (3) technologies that favor women; and (4) technologies that constitute social relations that are more equitable than those that were constituted by a prior technology or than those that prevail in the wider society. While I plan to use gender equity as my primary focus, I do not want to rule out any of the other candidates. Each gives us a way of thinking about the notion of feminist technology and allows us to note something important about the gender–technology connection.

STS THEMES AND THE FEMINIST TECHNOLOGY QUESTION

In the introduction to *Sex/Machine*, Hopkins (1998) lays out what he takes to be the four key themes expressed in the gender and technology literature. Adapting these themes to the feminist—rather than gender—technology connection yields the following four relationships: (1) Technology is associated with feminism; (2) Technology reinforces sexist (masculinist) social systems; (3) Technology subverts sexist (masculinist) social systems; and (4) Technology alters the very nature of gender and sex.

To address the fourth theme, Hopkins includes a section entitled "Body Building: The (Re)Construction of Sex and Sexuality" containing pieces on cosmetic surgery, surgery on intersexed infants, and adult sex change operations (transsexism). Sex and gender alterations raise the deepest questions at the core of feminism. If the goal of feminism is gender equity, then should equity be sought by deemphasizing gender differentiation and disconnecting social expectations with sex and gender? Or should we hold on to the idea that there are significant differences between men and women and insist that there should be equality of a kind that recognizes difference? On the face of it at least, it would seem that cosmetic surgery and sex-change operations accept gender differentiation; they accept sex as binary—one is either male or female. The technology, then, provides individuals with a means to fitting or better fitting into one of these categories. If anything, this seems to promote gender differentiation. Thus, if the goal of feminism is to eliminate or at least diminish gender differentiation, sex and gender alteration may be nonfeminist or antifeminist. On the other hand, feminists might argue that sex alteration creates a more complicated array of sex and gender categories and in this way challenges the idea of sex as binary. The point is still that different accounts of feminist goals lead to quite different perspectives on sex- and gender-altering technologies. Thus, although sex- and gender-altering technologies are important sites for study, they are not easily brought into our pursuit of the feminist technology question.

I have already discussed the association between technology and masculinity and the association of certain technologies with men and others with women. Technologies also can, and have been, associated with feminism. For example, feminists have fought for a variety of forms of birth control and, thus, many associate birth control with "women's liberation" and, hence, with feminism. Of course, whether the particular forms of birth control we have now are good for women, bring about equitable social relations, or favor women are different matters (see Hardon, and Aengst and Layne, this volume). In other words, just because certain technologies may be associated with feminism does not make them feminist.

The problem is that association is much too weak a relationship to ground a notion of feminist technology. Associations can be wholly inaccurate and can be deliberately shaped by advertisers, policy makers, and others who have non-feminist or even antifeminist interests. Take, for example, the construction of Virginia Slims as feminist cigarettes (Craig and Moellinger 2001). Advertisers of Virginia Slims sought to convince women that smoking this particular brand of cigarette was a sign of their freedom and liberation. However, while these cigarettes, deliberately packaged and advertised for women, made it more socially acceptable for women to smoke, they marked women as different from men, and played on the troubled association between femininity and slimness. Moreover, they promoted something that was detrimental to women's health, undermining women's better longevity. So, although we can understand how the association between feminism and Virginia Slims could be constructed, Virginia Slims hardly seems a good candidate for a feminist technology.

This is not to deny the importance of associations but rather to set them aside as the foundation for a notion of feminist technology. The question of feminist technology should be concerned with something more; it should ask not just about cultural associations but also about how technologies actually and potentially might affect the gender system and the circumstances of women.

REINFORCEMENT, SUBVERSION, CONSTITUTION OF A GENDER SYSTEM

I have adapted Hopkins's second and third themes (that technology reinforces gender systems and that technology subverts gender systems) for the feminist–

technology connection as themes of technology reinforcing sexist (masculinist) social systems and technology subverting sexist (masculinist) social systems. These provide better starting places for addressing the feminist technology question—although they only serve as a starting place. Both themes focus on social relations (a gender system) and then identify a particular way in which technology affects social relations; that is, reinforcing or subverting.

Notice, however, that neither of these themes provides a picture of how better, more equitable, or feminist social relations would be constituted. On the contrary, if we assume that most societies have gender-inequitable social relations, then technologies that reinforce a gender system will reinforce gender-inequitable systems. Of course, technologies that subvert will subvert a gender-inequitable system, but subversion doesn't guarantee that a more equitable system emerges. Rather, the reinforcement-subversion themes point to a relativistic notion of feminist technology. Both themes presume a prevailing gender arrangement, one that is strengthened (reinforced) or weakened (subverted) by technology.[2] By pointing backward to existing social arrangements, these themes suggest that in looking for a feminist technology, we should ask whether a technology is keeping things the same or undermining the prevailing order.

Hopkins has, it should be noted, only identified two possibilities here when there are probably more. A new technology could not just reinforce gendered social relations, but it might also strengthen gendered social relations in such a way that men have more power than they had in the past and women have less. Likewise, a technology might subvert gendered social relations by making inequities worse rather than better. Two quite different points need to be made in terms of the reinforcement and subversion themes. The first is that reinforcement and subversion don't cover the entire landscape, and the second is that neither reinforcement nor subversion provides us with an understanding of what constitutes improved (or any other form of feminist) social relations.

Two classic cases in the gender and technology literature are worth discussing here. Cockburn and Ormrod (1993) provide an analysis of the development of a microwave oven in which prevailing notions of gender become reinforced and reinscribed in the design of a new oven. Here reinforcement leads to the antithesis of a feminist technology. The design process is gendered; that is, it has gendered social arrangements and a gendered institutional culture. These gender relations are, then, reinscribed in what is produced, a new oven. The new microwave oven reinforced prevailing ideas about men and women and their roles in the kitchen. Cockburn and Ormrod's study suggests that from the perspective of design and production of technology, we have a chicken-egg problem. We may have to have feminist social relations at all stages of the design and development process in order to produce feminist technologies. Interestingly, this chicken-egg problem is roughly the same one that Butler struggled with in *Gender Trouble* (1990): How can we conceptualize nongendered social relations when we think with concepts and language that are deeply gendered? Cockburn and Ormrod take this a step further by drawing our attention to technology as a component of both the chicken and the egg.

The second classic case that is helpful here is Weber's analysis (1997) of the redesign of an airplane cockpit. This case involves reinforcement, subversion, and re-constitution of gender relations. The redesign of the airplane cockpit took place against the backdrop of an initial specification that reinforced the long-standing practice of not allowing women to be pilots. The initial specifications for the airplane cockpit were based on tables that provided the range of bodily measurements of men. The cockpit was designed to accommodate 90 percent of men, leaving out the tallest and shortest 5 percent. This meant that 70 percent of women were too small to safely function in the cockpit. Weber gives an account of what happened when the secretary of defense ordered that women be allowed to compete for all aviation assignments, including combat. Military policy analysts were horrified that the existing design for a new training aircraft, one that all pilots would have to master before moving on, would exclude the majority of female candidates. A working group was formed and came up with a new specification that would accommodate 82 percent of the female population. Using the language of reinforcement-subversion-constitution, the old cockpit reinforced inequitable gender relations; the new cockpit subverted the old system; and the new cockpit helped to constitute a more equitable gender system of piloting. Whereas the old cockpit was an obstacle for women to become pilots, the new one helped make it possible.

Thus, the new cockpit seems a good candidate for a feminist technology. More important, however, the account helps us to better understand the tension about the materiality of artifacts and their role in sociotechnical systems. The materiality of the prior cockpit—its dimensions and design features—did not allow equitable gender relations. Although some women might have been able to fit into the cockpit, far fewer women could safely fit (as compared to the number and percentage of men who could fit). On the other hand, the new cockpit had material features that were compatible with more equitable gender relations. Thus, we see that the materiality of the artifact made a difference in gender equity.

Nevertheless, the new cockpit did not and could not automatically bring about gender equitable relations. (Artifacts do not determine social relations.) The dimensions of the new cockpit did not necessitate that women be the pilots; men could and did fit in the new cockpit. Other aspects of the sociotechnical system had to come into play before the number of women pilots would increase. The rules had to be changed, women had to want to be pilots, and, more generally, the network of social relations had to reconfigure to make for women pilots.

We might think of the material features of an artifact as necessary but not sufficient conditions for gender equitable relations. The materiality of the cockpit seems to affect the situation in this way. We can imagine situations in which the materiality of an artifact allows gender equity, but because of the social meaning of the situation and ideas about gender, what the artifact allows in the way of gender equity is not perceived or used. The new design of the cockpit accommodated nearly as many women's bodies as men's and eliminated a major rationalization for prohibiting women from being pilots, but there was no guarantee that the possibility of equity would be taken up. Other things had to happen to bring about a shift in who could and would be a pilot.

This is not the whole story, but thinking of the material features of an artifact as necessary but not sufficient conditions for gender-equitable relations gives us some leverage on thinking about our two feminist technology questions. We can now return to those two questions: (1) Are there (or could there ever be) artifacts that could be considered feminist? and (2) Are there (or could there ever be) sociotechnical systems that could be considered feminist?

THE TWO FEMINIST TECHNOLOGY QUESTIONS CONFRONTED

Feminist Artifacts

The airplane cockpit case suggested that particular kinds of artifacts—and by extension the constructed, material world in general—might be necessary although not sufficient conditions for gender-equitable social relations. Artifacts and the constructed world create possibilities and constrain other possibilities. Nevertheless, the materiality of artifacts does not alone determine social relations or sociotechnical systems. Once the cockpit was built so that a large percentage of women as well as men could function in it, the number of women who became pilots did not automatically become equal to the number of men pilots. Other components of the sociotechnical system had to fall in line. And, going in the other direction, before the cockpit came to be redesigned, certain gender relations had to have changed to bring about the pressure (wish, imperative) to redesign the cockpit more equitably.

This, of course, is not news to STS scholars. The field takes as its starting place a critique of technological determinism. When an artifact is created or introduced or used, its material features do not determine the social relations around it. Once we recognize this, any hope that we might have had of finding an artifact that would, through its materiality, compel equitable gender relations is gone; we have to let go of the possibility of that sort of account of feminist technology.

Nevertheless, the materiality of artifacts is extremely important because it can facilitate or constrain equitable gender relations. How the artifact constitutes gender relations (and others) is contingent in complex ways. Certain patterns of social relations may be maintained despite the fact that the artifacts make equity possible and, in the other direction, pressures for more equitable gender relations may push against the materiality or design features of artifacts (as in the cockpit design). For an illustration of the former we need only think of baby bottles and breast pumps that make it possible for men and women to share equally (though differently) the responsibilities of infant care. Yet despite the availability of these artifacts and the more equitable gender relations they make possible, patterns of unequal responsibility for infant care persist. For the latter, consider the first women to occupy traditionally male professional roles; the material environments in which they worked were often ill-designed for them, with no convenient bathrooms, chairs, tables, and equipment that fit women's bodies, and so on. The built environment was not conducive to women having such jobs.[3] Yet women often persevered—despite the built environment, despite the unequal infrastructure. They pushed against the material infrastructure and have had some success in transforming it; it was not determinative.

The nondeterminative role of artifacts is often identified with artifacts being malleable or having interpretive flexibility (Pinch and Bijker 1987). Of course, some artifacts are more malleable than others in their materiality and design features (Winner 1986). For example, computers are extremely malleable in what they can be made to do and how they can be used. On the other hand, airplanes and lamps may be less malleable. Whatever the degree, however, all artifacts are malleable in the sense that they are compatible with different sociotechnical systems. What an artifact "is" (what it is understood to be) is a function of its place in a sociotechnical system. It is precisely because of this interpretative flexibility that artifacts are not determinative.

Designing artifacts for gender equity cannot, then, determine equitable gender relations. Nevertheless, it is important not to jump to the conclusion that artifacts are irrelevant to feminism or feminist goals. Artifacts and the built environment are an important factor in shaping gender relations; they can facilitate equitable gender relations and they can make such relationships difficult to achieve. Artifacts and the built world make a difference in what sorts of social relations are possible, what sorts of relations are easy or difficult, and what sorts of social relations are formed.[4]

Before turning to sociotechnical systems, the discussion of artifacts helps to explain two issues that were raised earlier about feminist technology—whether we should consider feminist technology a relativistic notion and whether we should consider it to be about favoring women rather than just about equity. Many of the artifacts that are part of our world and much of our built environment have been constructed with and around unequal gender relations and, consequently, seem to favor men. Think here of medicines designed with men's bodies in mind, the size of chairs, the heights of tables, the size of tool handles, the placement of bathrooms, and so on. Given this backdrop—a historical condition—artifacts and the built environment will have to be changed to create a gender-equal material infrastructure. Thus, it makes good sense in this historical period to think of feminist technologies as those that move in the right direction even if not achieving full gender equity. It also makes good sense to change the artifactual infrastructure in ways that favor women because this could bring about balance overall; it could equalize what is otherwise an unequal infrastructure. A plausible account of feminist technology could, then, refer to new artifacts and changes to the built environment that create the possibility for more equitable gender relations, including artifacts that favor women.

Feminist Sociotechnical Systems

The analysis of feminist artifacts points to the importance of the sociotechnical systems framework. Feminist technologies will be those in which artifacts and social relations will work together to achieve gender-equitable arrangements. This is consistent with our starting place in recognizing that technology and feminism come together around social relations; only now we see that social relations are constituted in sociotechnical systems in which artifacts and social relations are intertwined. Whether we focus on gender equity, what is good for women, what improves upon prevailing gender relations, or that which favors women, we have to look at a system in which all the parts contribute to the functioning of the system. This, of course, only describes feminist technology; it does not provide criteria for identifying what is equitable or what is good or better for women or what favors women. This combined with the chicken-egg problem make it unlikely that we will be able to identify a technology (sociotechnical system) that is unambiguously feminist. Most existing sociotechnical systems are likely, to some extent at least, to reflect the gender relations of the past as well as those that prevail in the broader world of which the system is a part. Thus, they are not likely to constitute or be constituted by gender-equitable relations.

There are other problems with the concept of a sociotechnical system. While obviously complex, it is an especially fluid concept. What counts as a particular sociotechnical system—what is inside and what is outside a particular system—seems to be entirely a matter of how a researcher or viewer chooses to draw the lines. Thus, the gender relations of a sociotechnical system can be delimited in any number of ways. This is important because we are likely to find mixtures of gender relations of various kinds in complex systems; there will be the gender relations in the design and development, manufacturing and distribution, marketing, meaning, and use of a technology. How a sociotechnical system is delimited may, thus, skew what can be said about its gender relations and whether the system is feminist. A sociotechnical system may be gender equitable in some ways and not in others. It may have a mixture of feminist and nonfeminist or antifeminist features.

These points are illustrated by considering a large sociotechnical system such as the Internet. The Internet can be thought of as a single sociotechnical system or as a cluster of subsociotechnical systems. There are search engines, chat rooms, news services, Web sites of a wide variety of kinds, e-business activities, hackers, the open source movement; there are programs, quasi-regulatory organizations, computers, telecommunications lines, and satellites. Although no one seems to have claimed that the totality is feminist, there has been a good deal of discussion about whether and how the Internet helps, or even empowers, women (Frederick 1999). In the very early literature about the Internet, social theorists claimed that the Internet was beneficial to gender equity because the gender of the person with whom one communicates online is not obvious—or at least not obvious in the ways that gender comes into play in face-to-face interactions. The Internet environment was thought to be blind to physical characteristics as expressed in the cartoon that depicted a dog saying, "On the Internet, nobody knows you're a dog" (Steiner 1993). More recently, scholars have argued that the Internet helps women who are confined to the domestic domain; it opens up opportunities for exchange, for having a voice, for forming political alliances, and so on (Balka 1993). Of course, certain subsystems are far from gender blind. Facebook and MySpace are filled with pictures that delineate gender, as do Webcams.

My point is that whether and how the Internet affects gender social relations depends on which aspect of the Internet—which subsystem—one considers. Gender comes into play in how the Internet is used—think of the huge proportion of Web sites devoted to pornography. Gender comes into play in who is using the Internet and how. Gender is a factor in the culture and composition of the institutions that control and maintain the Internet and the institutions that supply software, hardware, Internet service, and so on. Thus, the relationships between the Internet and gender as well as the Internet and feminisms are all over the place. Just about any social phenomena that can be found offline can be found online. The Internet, as a single sociotechnical system, is much too big and complex to be thought of as feminist or nonfeminist.

Thus, to say that a sociotechnical system is or is not feminist requires particular attention to how the sociotechnical system is circumscribed. Even so, the analysis of any sociotechnical system is likely to reveal a mixed picture.[5] To be sure, sociotechnical systems are a key site for understanding the co-construction of technology and gender, but the important thing here is not so much deciding whether the system is feminist or not but understanding how the sociotechnical system constitutes gender relations. Given the chicken-egg problem referred to earlier, technologies that are good, better, or favor women and those that constitute gender-equitable social relations are most likely to be those that have been designed, developed, or evolved (through users) with women's involvement or at least with women's rights and interests in mind. Otherwise, it is too easy to slip into re-constituting the world the way it is.

CONCLUSION: FEMINIST TECHNOLOGY

At the beginning of this chapter, I broke the feminist technology question into two connected questions, one about artifacts and the other about sociotechnical systems. I included the first question to ensure that we not lose sight of the materiality of technology. My analysis has indicated that we will not find artifacts that alone determine feminist social relations, but we will find artifacts that help to constitute (or get in the way of constituting) feminist sociotechnical systems. In other words, artifacts can constitute feminist social relations in combination with the other components of a sociotechnical system. Most important, my analysis shows the importance of artifacts and the built world in constituting gender-equitable relations. Thus, artifacts and the built environment must be part of the feminist agenda because they make feminist goals harder or easier to achieve.

If artifacts can be feminist only as part of sociotechnical systems that constitute feminist social relations, can there be feminist sociotechnical systems? My analysis suggested two problems with giving an account of this notion. First is the problem of describing feminist social relations. We seem to be better at identifying inequities and nonfeminist arrangements than at figuring out the alternative; that is, how feminist social relations would be constituted. The second problem is that of delineating the sociotechnical system. Here my analysis suggested that sociotechnical systems may be circumscribed in ways that make it difficult to give a simple "yes–no" answer to the question whether the system is feminist. We are much more likely to find a mixed picture. This is, of course, not an obstacle but rather an alert to the complexity involved in pursuing the feminist technology question.

Finally, in trying to answer the feminist technology question, I put forward four different ways that we might think about feminist technologies. If we acknowledge the chicken-egg problem, this multiplicity of criteria may be the best we can do for now; that is, we may have to make do without a sharply delineated notion or set of criteria. The important thing is that technology stays in the sights of the feminist social movement and that feminists call for sociotechnical systems that are good for women, gender equitable, sometimes favor women, and are always an improvement over prior gender-inequitable social relations.

NOTES

I am grateful to the editors of this volume for bringing this topic to my attention and for their insightful comments on earlier drafts of this paper. Ongoing conversations with Linda Layne and Shirley Gorenstein over many years have been enormously helpful to my thinking on gender and technology.

1. In "Technologies of Autonomy," Vestby (1996) treats parenting as a sociotechnical system and shows how the telephone has become part of parental care and control, allowing for "remote parenting."

2. A slight distraction here is that, in theory at least, a prevailing gender system could involve feminist social relations and, if this were the case, a new technology that reinforced the prevailing pattern would be feminist. Likewise, a subverting or disrupting technology introduced into a feminist system could be antifeminist insofar as it subverted gender equable social relations.

3. Issues of this kind persist in the workplace. For example, at my own university, it has recently come to the fore that women who work late into the night in their labs—doing so is essential to their professional progress—are fearful of their environment. The buildings have little security and women must walk to their cars through very poorly lit spaces. Although safety has no doubt been a concern in the design of and improvements to the campus, the safety of women faculty has not, evidently, been taken into account.

4. Smeds et al. (1994) give an account of the adoption of the central vacuum cleaning system in Finland. While purchasing the system did not make a huge difference, they report that purchase was "followed by a more equal sharing of the task of vacuum cleaning between men and women in the home" (39).

5. See, for example, Martin's account of women's use of the telephone in the late nineteenth century. She concludes that the telephone had contradictory effects on women: "[I]t had some emancipatory influence, yet it often contributed to reproduction, and even reinforcement, of sexist attitudes" (1998, 70).

References

Balka, Ellen. 1993. Women's Access to On-Line Discussions about Feminism. Electronic Journal of Communication 3(1). http://www.cios.org/www/ejc/v3n193.htm (Accessed July 1, 2008.)

Bijker, Wiebe E. 1995. "Sociohistorical Technology Studies." In S. Jasanoff, G. E. Markle, J. C. Petersen, and T. Pinch, eds., *Handbook of Science and Technology Studies*, 229–56. London: Sage.

Butler, Judith. 1990. *Gender Trouble: Feminism and the Subversion of Identity.* New York: Routledge.

Callon, Michel. 1986. "Some Elements of a Sociology of Translation Domestication of the Scallops and the Fishermen of St. Brieux Bay." In J. Law, ed., *Power, Action and Belief: A New Sociology of Knowledge?*, 196–229. London: Routledge & Kegan Paul.

———, and Bruno Latour. 1992. "Don't Throw the Baby Out with the Bath School! A Reply to Collins and Yearly." In A. Pickering, ed., *Science as Practice and Culture*, 343–68. Chicago: University of Chicago Press.

Cockburn, Cynthia, and Ruza Furst-Dilic, eds. 1994. *Bringing Technology Home: Gender and Technology in Changing Europe.* Buckingham: Open University Press.

———, and Susan Ormrod. 1993. *Gender and Technology in the Making.* Thousand Oaks: Sage.

Collins, Harry M., and Steven Yearley. 1992. "Epistemological Chicken." In A. Pickering, ed., *Science as Practice and Culture*, 301–26. Chicago: University of Chicago Press.

Craig, Steve, and Terry Moellinger. 2001. "'So Rich, Mild, and Fresh': A Critical Look at TV Cigarette Commercials, 1948–1971." *Journal of Communication Inquiry* 25, no. 1: 55–71.

Frederick, Christine Ann Nguyen. 1999. "Feminist Rhetoric in Cyberspace: The Ethos of Feminist Usenet Newsgroups." *The Information Society* 15: 187–97.

Frye, Marilyn. 2000. "Feminism." In L. Code, ed., *Encyclopedia of Feminist Theories*, 195–97. London: Routledge.

Hopkins, Patrick D., ed. 1998. *Sex/Machine.* Bloomington: Indiana University Press.

Hughes, Thomas P. 1994. "Technological Momentum." In L. Marx and M. R. Smith, eds., *Does Technology Drive History? The Dilemma of Technological Determinism*, 101–14. Cambridge, Mass.: MIT Press.

Law, John. 1987. "Technology and Heterogeneous Engineering: The Case of Portuguese Expansion." In W. E. Bijker, T. P. Hughes, and T. Pinch, eds., *The Social Construction of Technological Systems*, 111–34. Cambridge, Mass.: MIT Press.

Lie, Merete, and Knut Sorensen, eds. 1996. *Making Technology Our Own?: Domesticating Technology into Everyday Life.* Oslo: Scandinavian University Press.

MacKenzie, Donald, and Judy Wajcman, eds. 1999. *The Social Shaping of Technology* (2nd ed). Buckingham: Open University Press.

Martin, Michèle. 1998. "The Culture of the Telephone." In Patrick Hopkins, ed., *Sex/Machine*, 50–74. Bloomington: Indiana University Press. Reprinted from Michèle Martin, "Hello Central?" *Gender, Technology, and Culture in the Formation of Telephone Systems.* Montreal: McGill-Queen's University Press, 1991.

Oblepias-Ramos, Lilia. 1998. "Does Technology Work for Women Too?" In Patrick Hopkins, ed., *Sex/Machine,* 89–94. Bloomington: Indiana University Press.

Pinch, Trevor J., and Wiebe E. Bijker. 1987. "The Social Construction of Facts and Artifacts: Or How the Sociology of Science and the Sociology of Technology Might Benefit Each Other." In W. E. Bijker, T. P. Hughes, and T. Pinch, eds., *The Social Construction of Technological Systems,* 17–50. Cambridge, Mass.: MIT Press.

Smeds, Riitta, Outi Huida, Elina Haavio-Mannila, and Kaisa Kauppinen-Toropainen. 1994. "Sweeping Away the Dust of Tradition: Vacuum Cleaning as a Site of Technical and Social Innovation." In C. Cockburn and R. Furst-Dilic, eds., *Bringing Technology Home: Gender and Technology in Changing Europe,* 22–41. Buckingham: Open University Press.

Stanley, Autumn. 1998. "Women Hold Up Two-Thirds of the Sky: Notes for a Revised History of Technology." In Patrick Hopkins, ed., *Sex/Machine,* 17–32. Bloomington: Indiana University Press. Reprinted from Joan Rothschild, *Machina Ex Dea: Feminist Perspectives on Technology.* New York: Pergamon Press, 1983.

Steiner, Peter. 1993. "On the Internet, Nobody Knows You're a Dog." *New Yorker* 69, no. 20: 61.

Vestby, Guri Mette. 1996. "Technologies of Autonomy." In M. Lie and K. Sorensen, eds., *Making Technology Our Own?: Domesticating Technology into Everyday Life,* 65–90. Oslo: Scandinavian University Press.

Wajcman, Judy. 2004. *TechnoFeminism.* Cambridge: Polity Press.

Weber, Rachel N. 1997. "Manufacturing Gender in Commercial and Military Cockpit Design." *Science, Technology and Human Values* 22, no. 2: 235–53.

Winner, Langdon. 1986. *The Whale and the Reactor.* Chicago: University of Chicago Press.

2 The Need to Bleed?

A Feminist Technology Assessment
of Menstrual-Suppressing Birth Control Pills

JENNIFER AENGST AND LINDA L. LAYNE

SEASONALE[1], A LOW-DOSE, extended-regimen, birth control pill approved by the U.S. Federal Drug Administration (FDA) in September 2003, regulates menstruation so that it occurs only four times a year. Seasonale shares the same chemical makeup as monthly birth control pills, so it shares with conventional pills a high level of contraceptive efficacy (99 percent if used consistently)[2], and a number of possible health benefits (a reduced risk of cancers of the ovaries and uterus, pelvic inflammatory disease, ovarian cysts, ectopic pregnancy, and noncancerous lumps or cysts of the breast) (Duramed 2005b). Like conventional pills, Seasonale also increases the risk of "blood clots, stroke, and heart attack," all of which are greater among smokers, and like other oral contraceptives, Seasonale "doesn't protect against HIV or other sexually transmitted diseases" (Duramed 2005a).[3]

The unique purported benefit of Seasonale is that it offers women "Fewer Periods. More Possibilities,"[4] and it is this aspect of the technology on which our analysis focuses. Several other birth control pills with a range of menstrual-suppressing features are also coming to market[5] including Seasonique (similar to Seasonale), Lybrel (which suppresses indefinitely), and Yaz and Loestrin24Fe, both of which are taken twenty-four, instead of twenty-one, days, followed by four "reminder" placebo pills, therefore offering shorter (three-day), lighter, monthly periods. Issues raised by Seasonale apply equally to these and other menstrual-suppressing birth control technologies.[6]

The subject of menstrual suppression has been debated in both medical and popular literature. Several of these articles make use of the responses of over 900 visitors to the online Museum of Menstruation and Women's Health (MUM) to the question, "Would you stop menstruating indefinitely—for years, maybe—if you could start up again easily if you wanted a child?," which has been posed on this Web site since 2000.[7] In this chapter, we too draw on these data, along with a small survey of Aengst's social network,[8] and a review of the writings of advocates and opponents to show how physical and attitudinal differences among women complicate the question of "feminist technologies."[9] We also consider how different feminist theories shape the evaluation of technologies for menstrual suppression as we address the meanings of "nature," "normality," and "necessity," gender norms, and the appropriation of technology. The result is what Layne calls a "feminist technology assessment" (see the introduction to this volume).

THE EXPERTS: FOR AND AGAINST MENSTRUAL SUPPRESSION

Women's health experts are widely divided on the issue of menstrual-suppressing birth control pills. The most common argument made in favor of suppression is that promoted by a Brazilian gynecologist, Elsimar Coutinho, and Sheldon J. Segal, an endocrinologist at the Population Council.[10] Coutinho did research in 1959 at the Rockefeller Institute for Medical Research with Dr. George Corner, one of the codiscoverers of progesterone. After returning to Brazil he conducted a clinical trial of Depo Provera. The drug was being tested for the prevention of threatened spontaneous abortion and premature delivery, at which it failed, but the trial inadvertently revealed the drug's capacity to suppress ovulation and menstruation. He then conducted a series of clinical trials for the maker, Upjohn, on volunteers who did not want to become pregnant, which showed how Depo Provera could prevent ovulation and menstruation for one, three, or six months depending on the dose (Coutinho and Segal 1999, 9). This side effect, referred to in the women's health movement as a "menstrual disturbance" (Hardon, this volume) is lauded by Coutinho and Segal as a marketable benefit of the drug.

The argument put forward in their book is one that had been articulated nearly twenty years earlier by Dr. Barbara Harrell (1981, 816): menstruation is "relatively uncommon" for preindustrial women because of frequent pregnancies and prolonged lactation resulting in amenorrhea and that "continuous menstrual cycling is not a natural attribute of human females." Coutinho and Segal (1999, 2) go further and suggest that "repeated menstruation" could be harmful to women's health (1999, 5) and that suppressing menstruation would have health benefits for "women who suffer from anemia, endometriosis, or PMS."[11]

They end their book celebrating the liberatory feminist potential of this drug. They envision a future in which women use menstrual-suppressing drugs, and as they grow "more confident, would lengthen the menstruation-free interval . . . other women would be encouraged to try" and medical researchers would be motivated to find more advanced methods. . . . This would forge a major advance in women's health, led by women. The pioneer feminist Margaret Sanger wrote,

'No woman is completely free unless she has control over her own reproductive system.' Let this new freedom begin" (1999, 1603–4).

Other supporters, Charlotte Ellertson, Ph.D. (also of the Population Council), and Sarah Thomas, B.A.,[12] argue that the "view that menses, even debilitating ones, are normal, stigmatizes menstrual disorders and deprives millions of women of legitimate and easily available help" (Thomas and Ellertson 2000, 924). Using the liberal feminist tropes of personal control, choice, and liberation, they construe menstrual suppression as feminist, asserting that the use of contraceptives such as Seasonale "lets women *control* their hormonal profiles as well as whether and when they *choose* to bleed" and that menstrual suppression will "contribute to happier, *less encumbered lives*" (2000, 922, emphasis added).[13] They situate menstrual suppression as just one of several choices women routinely make about menstrual and birth control products and in so doing suggest that the decision to suppress menstruation is a simple consumer choice, no different than deciding which type of sanitary product to buy (2000, 923).[14]

In the twenty-two American and Canadian articles published in the popular press between the 1999 publication of Coutinho and Segal's book and the FDA approval of Seasonale in 2003, advocates of menstrual suppression like these were quoted "twice as often as opponents" (Johnston-Robledo et al. 2006) and in 40 to 50 percent of the articles (2006, 357), proponents praised the way this drug "expand[ed] women's choices" and gave women "more control over their lives, menstrual cycles or reproductive health."

Those against argue that there are uncertainties about the effects of long-term usage, and note that menstrual-suppressing drugs like Seasonale reinscribe negative attitudes toward women's bodily processes (Prior and Hitchcock 2006).[15] Christine Hitchcock, a research associate at the Center for Menstrual Cycle and Ovulation Research, points out that the same hormones that create menstrual cycles "act in the brain, bones and skin" (Saul 2007, C4) and so altering them may contribute to unknown long-term health risks for women. She also worries about "the idea that you can turn your body on and off like a tap" (Saul 2007, A1). The Society for Menstrual Cycle Research issued a position statement on menstrual suppression in 2003 that cautioned that more research was needed regarding not only the medical consequences of use but also the psychosocial dimensions of suppression before women could make informed decisions. The authors recognized "that menstrual suppression may be a useful option for women with severe menstrual cycle problems such as endometriosis," but argued against use that would suppress "normal, healthy menstrual cycle[s]" and expressed special concern for the use of extended oral contraceptives by adolescents.

Opponents also point out that Seasonale does not deliver on the purported "convenience of four periods a year." During the first year of use, women are "more likely to have breakthrough bleeding (which varies from slight spotting to a flow much like a regular period) than with a 28-day birth control pill" (Duramed 2005b). In fact, during the first year, "total bleeding days are similar to a traditional birth control pill" (Duramed 2005b). They also point out that skipping the conventional monthly placebo pills puts users more at risk for breast cancer, heart attacks, and strokes. A further negative is the expense—one package of a single cycle of Seasonale (ninety-one tablets) ranges from $100 to over $200 (generic or brand name), which may or may not be covered by insurance. This is in contrast to monthly birth control pills that typically range in cost between twenty to thirty dollars per month, the equivalent of sixty to ninety dollars for three months.[16]

Opponents also voice concern about use by teens[17] who are perceived to be especially vulnerable to marketing efforts because they are more likely to have negative attitudes toward menstruation (Johnston-Robledo et al. 2006, 354), be uncomfortable with their bodies, and particularly concerned about issues of personal hygiene and the scrutiny and judgment of others. Furthermore, since many teens are only sexually active intermittently, the health risks of continuous birth control pills are considered an unnecessary risk. Contributors to the MUM Web site mention special concern for this population and observed that they themselves probably would have welcomed the chance to suppress when they were young, but now that they are older/wiser, see the harm in it.

Underlying this debate are conflicting interpretations of both Seasonale's and women's relationship to nature. Women have long been culturally associated with nature, and feminists have been, and continue to be, deeply divided about whether this association is beneficial or detrimental for women's status. Seasonale thus makes a particularly good case for considering the notion of "feminist technology" through the lenses of diverse feminist theories.

IS MENSTRUATION NATURAL?

Proponents of suppression challenge the naturalness of monthly menstruation. They suggest that the view that menstruation is natural is simply a myth (Thomas and Ellertson 2000, 922), and urge us to reexamine "the credo that frequent and prolonged menstruation is the 'natural' state." Other times, they agree with opponents that menstruation is natural but differ with them on the meaning and value attributed to this. Whereas opponents to suppression believe that menstruation is natural and thus should not be tampered with, supporters of suppression scorn such views. For instance Coutinho and Segal criticize those who "adopt the view that since it is 'natural' for women to menstruate, it must be good for their health. They seem to believe that 'you can't fool Mother Nature!' The logic is that things natural, such as pain, physical or mental impairment, or even disease should be accepted simply because they are natural" (1999, 138).

Similarly, Thomas and Ellertson cast menstruation as natural but treat the natural as something we can and often should change: "Health professionals and women ought to view menstruation as they would any other naturally occurring but frequently undesirable condition"; that is, by "eliminating" it.[18] They argue that "[s]uppression should be just one option for women and those who choose not to avail themselves of it certainly deserve to have their choices respected[19] . . . just as do women who choose to use 'natural' family planning methods in place of hormonal ones or 'natural' menstrual sponges in place of commercial tampons or sanitary pads" (2000, 923). In this passage they treat menstruation as natural but destabilize "natural" by putting it in quotes, liken menstruating (which most American women do) to the use of natural birth control methods and natural menstrual sponges (which most American women do not use), and suggest choosing not to take a drug to suppress "natural" monthly menses is something only a small portion of women would embrace.

MIMICKING NATURE

Both traditional birth control pills and Seasonale create planned periods that deviate from the normal menstrual cycle while appearing to mimic nature. With traditional birth control pills, women take seven days of placebo sugar pills after twenty-one days of oral contraceptives, which allows them to mimic the "normal"—that is, "natural"—monthly cycle. Seasonale users take the placebo pills after eighty-four days. The substitution of four so-called seasonal periods with Seasonale (instead of, say, five a year), is also clearly designed to appear "natural."

To counter the possible perception of unnaturalness of four periods, the makers of Seasonale stress that the monthly periods with traditional birth control pills are not in fact "real," "natural," or "normal" periods. As a promotional brochure for Seasonale explains, "when you take the Pill, you don't ovulate. This means your ovaries don't release an egg, the lining of your uterus doesn't build up, and you don't get a menstrual period. Instead, you get a 'Pill period'" (Duramed 2005b). Women using the pill, whether traditional or extended-regimen, simply "appear to menstruate."[20] The seven-day placebo week was historically included because of cultural notions of "normalcy." According to Carolyn Westhoff, M.D., professor of obstetrics and gynecology at Columbia University, "It was thought that women would find it reassuring to get a period every month. The week off was inserted not for biological reasons, but just to make women and doctors more comfortable" (Davis 2003).

In a *New Yorker* profile of John Rock, the devout Catholic physician who was one of the inventors of the birth control pill, Gladwell explains that one of the reasons Rock believed that the pill would be acceptable to the church is that it "was a 'natural' method of birth control . . . Progestin . . . is nature's contraceptive. And what was the Pill? Progestin in tablet form" (Gladwell 2000, 2).[21] Furthermore, Pope Pius XII had sanctioned the rhythm method in 1951 because he deemed it a "natural" method of regulating procreation, and Rock saw the pill as an extension of this. He "insisted on a twenty-eight-day cycle for his pill" in order to preserve the natural "menstrual rhythms" (Gladwell 2000).

IS MENSTRUATION NECESSARY?

Proponents of menstrual suppression argue that menstruation is unnecessary. According to Dr. Mitchell Creinin, director of family planning in the Obstetrics and Gynecology Department of the University of Pittsburgh's Magee-Women's Hospital, "The idea that a woman 'needs' to have a period is folklore" (Shaw 2003). Coutinho and Segal (1999, 159) assert, "Recurrent menstruation is unnecessary. . . . It is a needless loss of blood."

Thomas and Ellertson (2000, 923) draw analogies with other conditions now commonly controlled with pharmaceuticals, noting that "modern medicine is all about the artificial control of conditions that range from the life threatening, debilitating, and uncomfortable to matters of mere taste." An apt comparison might be made with aging. As with menstruation, some now see aging as a biological process that is "neither natural nor inevitable" (www.antiagingny .com). Groups like the New York City–based PhysioAge Medical Group purport to "stop" or "slow" the aging process through the use of hormonal replacement therapies (HRT), which had initially been targeted to women for menopause. They believe that they can "correct hormonal imbalances" so as to maintain youth longer and to prevent diseases associated with aging such as osteoporosis, cardiovascular disease, and Alzheimer's.[22] Both hormonal menstrual suppression and anti-aging HRTs illustrate how bodily processes that are considered "undignified" and detrimental to one's sexual appeal are, at least by some, no longer thought to be necessary.

MENSTRUAL SUPPRESSION:
AN ENHANCEMENT TECHNOLOGY?

"Enhancement technologies" have been described as those aimed at "improving human characteristics, including appearance and mental or physical functioning, often beyond what is 'normal' or necessary for life and well-being" (Hogle 2005, 695). These new, body-altering techniques repair, replace, and even redesign the human body in response to individual wants (Hogle 2005, 696). Well-known examples include the use of anabolic steroids for athletic performance enhancement, hormones for rejuvenation, and the use of human growth hormone for short children because height is thought to improve an individual's chances for success. These culturally shaped needs and desires reinforce already existing gender norms. For example, drugs are now sometimes being used to limit the height of tall girls who would otherwise exceed the standards of femininity. Are menstrual-suppressing birth control pills performance-enhancing drugs? Do they improve an individual's chances for success, and, if so, how?

Just as with soldiers, truck drivers, and test takers who are being given or self-administering drugs to enhance performance, the demands for productivity in the workplace may be legitimating hormonal menstrual suppression. For example, ob-gyn Shari Brasner explained how she began adapting normal birth control pills to suppress her periods (by skipping the placebo pills) when she "decided that my busy schedule really precluded the ability to take a break to go to the bathroom every couple of hours to take care of personal needs" (Harris and Saul 2006). Adds Linda C. Andrist, a professor at Massachusetts General Hospital's Institute of Health Professions in Boston, "[W]e don't want to confront our bodily functions anymore. *We're too busy*" (Saul 2007, A1, emphasis added).

Proponents of suppression often highlight the economic consequences of menstruation. For example, Andrew Kaunitz, a gynecologist who was one of the site-testers for Seasonale, notes that "[m]enstrual disorders represent a major cause of absenteeism from work" (Chesler 2006). A Canadian study that found that women afflicted by heavy menstrual bleeding give up $1,692 a year in lost wages (Saul 2007, C4). According to Thomas and Ellertson, "menstrual disorders cost U.S. industry about 8 percent of its total wage bill" (2000, 922). With gendered attitudes about work productivity, women's absenteeism is often interpreted as another example of how women's reproductive processes (menstrual cramps, pregnancy, child care) interfere with the efficiency of the workplace. A number of contributors to the MUM Web site mention how the economic system does not accommodate women's bodily needs. For example, one woman writes,

"I don't judge anyone who wishes to stop their menstruation, but I think that modern western existence is fundamentally anti-feminine, and that we are being reshaped into suffering worker drones for capitalism." Another woman writes, "In this world of male corporate culture, where most women work outside the home, it is difficult to take time for oneself as a woman without feeling like a 'whiner' and 'complainer.'"

In addition to its potential for enhancing productivity, menstrual suppression might be considered an enhancement technology along the lines of cosmetic surgeries intended to improve appearance and increase sexual attractiveness, given the fact that taboos against having sex while menstruating appear to still be diminishing women's sex lives.[23] Writing in 1976, Delaney, Lupton, and Toth devote a chapter, entitled "'Not Tonight, Dear,'" to the subject. They observe, "What is so remarkable about the sex taboos against menstruating women is that they have not faded into vestigial reminders of a primitive past; they are still very much a part of everyday life for most people" (1976, 14).[24]

Submissions to the MUM Web site indicate that the taboo is still a factor for some women. Some don't have sex during their periods either because of their own beliefs or those of their sexual partners, while others do but recognize that others might disapprove.[25] "I would gladly have sex while on my period, as I find it really does tend to lessen cramps, provided my partner wasn't disgusted by the whole affair." Another woman who would like to suppress her periods writes, "I don't like having my husband be so disgusted by my menstrual fluid that he will not have sex with me during that week. That hurts my feelings, and the worst part is it's not his fault. He was raised by women who were disgusted by their own bodies, and through them he was taught to be disgusted." Another confesses she "even ha[s] sex during that time and that always makes me feel better. Disgusting? Well, I don't care. It's not to me and it's not disgusting to my husband. It's just the normal me. You can say what you want but I have the feeling it's a lot more fun the way I do it instead of . . . pushing my husband out of bed." One woman advises, "Have sex if you want to when you're bleeding—men need to get over their fear of blood—it's only your mind holding you back. I have been blessed with men in my life that have no issues with menstruation and sex."[26]

SCHEDULED MENSTRUATION: THE MODERN, YET FEMININE ALTERNATIVE?

One important feature of Seasonale is that it allows women to schedule their periods. In this regard, menstruating is comparable with other reproductive events that American women are increasingly likely to schedule, including birth (either via scheduled induction or scheduled caesarian sections) and pregnancy loss. Induced labor has become much more common in the United States in the past fifteen years. The rate of inductions doubled, growing from 9.5 percent in 1990 to 20.3 percent in 2003.[27] This trend is seen as "a product of our times" (Fink 2000) that suits "our fast-paced lives" and allows women to exercise their "choice" and ensure that the baby not "arrive at an inconvenient time" (Lane 2006). In this way, Lane reports, families can plan on being home for the holidays, be sure the baby is born when the family is all together (e.g., before the father is shipped to Iraq), and accommodate the "career demands of both parents" (Lane 2006). It also spares women the "discomforts of late pregnancy" (Lane 2006). Similarly, the rate of elective caesareans has increased in the United States. In 2006, 31 percent of births were c-sections, up 50 percent from 1996 (Park 2008). Estimates of how many of these were elected are as high as 18 percent (Park 2008).[28] The growth is attributed in part to a growing number of women who are requesting elective caesarians. The reasons given are for "convenience—the ability to fit childbirth into their work schedules, plan for the care of their other children, or have spouses, parents, or both present at the birth" (Brody 2003).[29] Others prefer it for cosmetic reasons, to avoid the "vaginal stretching and mauling" of natural birth (Hamer 2007a). Anecdotally, there is evidence that because of the tax benefits of having a dependent, the planned induction and c-section rates are highest in December.[30]

Layne (2003, 4) describes having used injections of the hormone progesterone to postpone an imminent miscarriage until after a professional conference, so as not to add a professional loss to her personal one. Although this practice may be rare, many women with diagnosed fetal demise choose to schedule a D&C or induction, depending how far along they are, rather than waiting for "nature to take her course."

The ability to schedule one's periods is being pitched as a sign of modern womanhood. In their book, *Is Menstruation Obsolete?*, Coutinho and Segal argue that menstrual suppression is a distinctly "modern" solution for the problems of modernity (including unnaturally frequent menses), because it is modern life that has led women to deviate from what the authors believe is their "natural" state of continual pregnancies and breast-feeding. Similarly, Dr. Leslie Miller, associate professor of obstetrics and gynecology at the University of Washington, is quoted as saying, "[S]uppression of menstrual cycles is a modern solution to a modern lifestyle" (www.noperiod.com).[31]

On the Seasonale/Seasonique Web site (2007), women are invited to use the "Personal Planner" to schedule "events like vacations, business travel, romantic encounters, and family reunions based on your inactive Pill dates," and a write-up about Seasonale on the Cleveland Clinic's Health Information Center Web site assures, "[O]n a wedding day, honeymoon, or family vacation, for example, no woman wants to have the added burden of menstruation."[32] The examples typify femininity, with honeymoons and weddings also signifying normative, heterosexual sexual activity. The type of woman who is envisioned in Seasonale's marketing campaign is young, middle or upper middle class,[33] works outside of the home, is heterosexual, and is a "modern" yet feminine woman. Modernity is invoked as a way to differentiate Seasonale from "traditional" birth control pills. An early advertisement shows women choosing one of a group of items (airport chairs, high-top sneakers, yoga mats) that are colored gray, and selecting the one that is pink. A more recent brochure develops the same trope, using the contrast between black and pink to promote the desirability of this product (Duramed 2005b). An attractive, young, feminine (skirt- and high-heel-wearing) woman is shown walking by a long row of identical black armchairs glancing back over her shoulder at the one pink chair. On the next page she is pictured sitting on the pink chair; that is, having chosen it and made it her own through the deployment of her body. On another double-page spread at the center of the brochure she is pictured looking down, admiring the pretty pink shoes she is wearing that she has clearly chosen out of an endless line of identical black ones. In addition to signaling femininity with pink, Seasonale uses the various versions of this visual metaphor to distinguish itself from boring, old-fashioned, ordinary birth control pills, and at the same time suggests that the consumer can distinguish herself as a young, attractive, feminine, fun-loving woman by choosing this form of birth control. The metaphor can also be read as referring to the way one's period days are different from ordinary days. In this reading, the occasional period is a fun, feminine alternative to ordinary, undifferentiated days. And with Seasonale, the pink days will be even more special because they are more rare and, like the pink shoes, an expression of personal choice. This trope also suggests the ease of use: chairs and shoes are familiar technologies that women already know how to use to accessorize their bodies

SEASONALE: THE MAINSTREAMING OF AN APPROPRIATED TECHNOLOGY

Seasonale presents an interesting example of "appropriated technology."[34] Birth control pills have long been adapted by users to schedule or eliminate their periods. Users simply skip the placebo pills of their monthly cycle. In their 1976 book, *The Curse: A Cultural History of Menstruation*, Delaney, Lupton, and

Extended-regimen SEASONALE®
The daily birth control pill that's the same but different.

seasonale

Extended-regimen SEASONALE® is more like traditional birth control pills than you might think

The protection and hormones are the same. The serious risks are the same. Even the once-daily routine is the same. But SEASONALE® does something a little different.

Learn more about it. Inside, you'll find out

- How SEASONALE® works
- Why you can have fewer periods
- Important safety information

seasonale   SEASONALE® tablets are indicated for the prevention of pregnancy.   Please see accompanying important Product Information.

Ask your healthcare professional if the difference is right for you

The extended regimen of SEASONALE® is a small departure from the regimen of your traditional Pill, but it can lead to some different possibilities.

Only your healthcare professional can tell you if SEASONALE® is right for you. So if you like the possibilities of an extended regimen, make an appointment today to talk about SEASONALE®. Because sometimes a little difference is the difference you should ask about.

Toth devote a chapter to various techniques women have used in "escaping the monthlies," among which they single out hormones as "undoubtedly the least unsafe suppressor" (1976, 215). Hormones have been given as menstrual suppressors to "paraplegic and severely handicapped women and to women in plaster casts," used by prostitutes to stop their periods indefinitely, and by athletes to delay a period until after an important athletic competition although women have won "gold medals and established new world records in the . . . Olympics during all phases of the menstrual cycle" (1976, 57). There is no accurate data on how many women are now regularly skipping placebo pills, but medical organizations and reproductive health centers—such as Reproductive Health Technologies Project and Association of Reproductive Health Professionals—acknowledge the practice (Thomas and Ellertson 2000). One contributor to the MUM Web site, a twenty-three-year-old biologist from Malaysia, reports that "dancers, athletes, and other women who find periods inconvenient have known about this trick for a long time" and she recalls an experience from her childhood when they "were going to the beach for a camp with their church and her twelve-year-old sister and her best friend were on their periods" and the friend's father, a gynecologist, "gave them some pills to stop their periods so they could swim." Gottlieb (2002, 388) reports that some Balinese women take birth control pills in order to delay the onset of their periods, "precisely timing the menstrual cycle" so that they can "participate in traditional temple rituals from which menstruating women are still actively banned." She interprets this "cultural conservatism" as being, like the veil, an expression of "ethnic pride and nationalism in the face of international pressure to Westernize" (Gottlieb 2002, 388).

"Appropriated technologies" represent a reversal of the typical power flow in technology design and production from those with high social power to consumers who may be outside the centers of social power, and thus may incline us to be positively predisposed to them. Indeed, we might interpret Seasonale as an example of "the collective force of [women] in shaping technology design through market demands" (Eglash 2004, xvi). But appropriated technologies are not necessarily liberatory. As Eglash (2004, xvii) observes, "[A]ppropriated technologies do not have an inherent ethical advantage. First, insofar as appropriation is a response to marginalization, we should work at obviating the need for it by empowering the marginalized. Second, not all forms of resistance are necessarily beneficial in the long run." Furthermore, Seasonale and the other "me-too" menstrual-suppressing birth control pills coming to market are simply packaging this already-appropriated technology and selling it back to women at inflated prices.

MENSTRUATION = WOMANHOOD?

Underlying much of the discussion surrounding menstrual-suppressing birth control are essentialist claims about femaleness. For example, psychiatrist Dr. Susan Rako, author of No More Periods? The Risks of Menstrual Suppression (2003), maintains that these new technologies are "doing away with women's normal hormonal menstrual cycle, which is really responsible for what fundamentally makes a woman a woman" (Cox and Feig 2003).[35] The respondents in Aengst's survey who were reluctant to use Seasonale linked a monthly cycle with femininity. Many of the contributors to the MUM Web site expressed similar views. One thirteen-year-old reports that although she "HATEs period pains" she "always feels blessed when 'it's that time of the month.' I love the feel of being a woman . . . I . . . like that I am growing up and . . . it makes me really happy and proud to be a woman." A thirty-six-year-old woman says she would never give up her period. "I like that it is regular, I love to feel the cramps. It reminds me of being in labor and the power that I found there. The power that resides deep within me, connecting me to all women who have lived before, and all to come." She is teaching her daughters by example that "cramps are part of the power of being a woman." A thirty-two-year-old U.S. mother of three who is a graduate student in Southern California and "would never give up [her] periods" says, "I am the woman I am in a large part because of my relationship with my body—my awareness of my cycle, my knowledge of how my parts work, my connection to my fertility." Another contributor writes, "To suppress menstruation is to suppress being a woman."

A woman from Zambia notes that even though it's inconvenient when traveling, "it's a wonderful experience of womanhood. It makes us different from men." Another writes, "[I]t's what makes us special." Several refer to their periods as "a gift." "You bleed each month because each month you have the potential to create LIFE. Screw being envious of men. We create men! We can make men and women right in our bodies and it is a beautiful and amazing privilege!" Another says, "[O]ur periods are like our trademark," and counsels that others should "be proud to be a woman." A forty-eight-year-old woman explains why she would not choose menstrual suppression: "I feel connected to other women around me and throughout time." A thirteen-year-old says, "I feel like it is a bond with all women, one of my few assurances that I'm normal. It assures me that I am healthy and similar to half of the earth's population."

Many others react to these expressed views. Of women who say that "they didn't feel like women without their periods," one mother replies "Bah!" and shares how much she would welcome "never having to deal with a bloody tampon again." A woman who has menstruated for thirty-five years says that she has "appreciated the feminine, the moon cycles, the fecundity, the fertility. It's awesome, . . . now go away, shoo. I am woman, hear me roar." One woman who reports she "would stop in a heartbeat if [she] could" explains, "I have been a teenager bleeding, a young woman bleeding, and a mother bleeding. Now I'm tired of bleeding. My womanhood has been proven!" A nineteen-year-old with very difficult periods replied, "I love being a woman and I feel empowered because of who I am, but my period does not make me a female. I don't need my period to remind me 'oh yeah, I am female.'" A forty-three-year-old from the United States says, "[T]his doesn't make me feel like a woman. It makes me feel dirty, like hiding all day. I feel like a woman when I can put on a pretty dress, not worry it'll get stained, and be intimate with my husband." Another writes that menstruation does not "prove your femininity. Hell, if I want to get in touch with my feminine side I'll look in my heart and mind, not at the red blood in the toilet."

MENSTRUATION = WOMYN/NATURE?

Many of the arguments against menstrual suppression not only allude to the special bond that menstruating creates among women, but also to the bond it creates between women and nature. One of the respondents to Aengst's survey commented, "I think the monthly cycle represents a powerful connection with the cycles of nature—a reminder that life is constantly in flux and corporeal bodies aren't the same at all times." A contributor to the MUM Web site who self-describes as "a pagan" uses her "blood in rituals" and likes "how it connects me with the earth, especially in our modern world." Another contributor writes, "I almost always get my period right around or after the full moon, . . . and I like that vague connection to the moon/the universe." A twenty-eight-year-old from Alabama states, "[T]here's no way that I would give up my Moon cycle other than for pregnancy or naturally occurring menopause!" A thirty-four-year-old from Virginia writes, "I enjoy the regular reminder of my power as a womyn and of my connection to the moon and the tides." And a woman in her early thirties says that although "at an earlier stage of my life I would have said yes" to suppression, "I learned later in my life that our moon cycle is a gift from the Goddess. . . . In ancient times women were revered and respected because she could bleed without a wound. The blood was given to Mother Earth to nourish her." Another writes, "Menstrual blood has been used to fertilize plant life (I give it to my plants—they love it!)" and one urges other women "to engage with our connection to the planet and stop harmful . . . activities. I wish every woman could observe nature on a daily basis, sit in her garden and tend vegetables, have time to make a simple meal, be able to sit, chat, sew and comfort."

Some object to the views of those who link menstruation with nature/womanhood. For example, one writes, "Those moon-womyn with their raspberry leaf tea just make me tired. Menstruating smells. Get over it. There's an odor and it isn't raspberry tea, sweetheart." Another contributor quips, "I think all these folks going on about their 'Moon Time' are full of it. They have never had a painful period." A more tolerant response is expressed by a woman who states, "I'm happy for women who feel that having their period connects them with the moon and the tides. For me, though, all that my period connects me with is a 500-count bottle of Advil and a heating pad." And a fifty-year-old who gets migraines before her period starts and has gotten her periods "regular as clockwork since I was 12" writes, "I don't need this any more. Mother Nature, lay off already."

IS SEASONALE A FEMINIST TECHNOLOGY?

Different feminist theories render different answers to this question. In this section, we begin by using the "how would this look if it were men" technique to begin our feminism assessment. We then consider the issue through the lenses of liberal, radical, socialist, essentialist/cultural, eco-, African American, existential, and cyborg feminism[36] and conclude with our own evaluations. Readers should keep in mind that we are describing feminist theories and that feminists often embrace more than one of these perspectives.

THE "HOW WOULD THIS LOOK IF IT WERE MEN?" TEST

One technique often used in thinking about whether a technology or social arrangement is feminist (or sexist) is to consider how the issue would look if the sexes were switched. In the case of menstrual-suppressing drugs, perhaps a comparable male case would be a semen- or ejaculation-suppressing drug. This comparison is in fact made by a twenty-six-year-old Portuguese contributor. She writes, "Look at men—do you think they see their sperm as repulsive? Oh, God, no! They tend to be proud of it. They're proud of their sexuality. . . . And do you think a period is more repulsive than sperm? Well, I personally don't think so."[37]

As this woman points out, like menstrual blood, ejaculate could be considered dirty, messy, ritually polluting, and/or inconvenient, and, like menstrual suppression, ejaculate suppression could serve as a method of birth control. Even though several drugs suppress ejaculation (e.g., Flomax, a drug prescribed to men who have enlarged prostates), this is not being marketed as a sales point but as an unwanted side effect.[38]

Liberal Feminism

Liberal feminism has tended to celebrate the expansion of choices for women, including increasing the number of options that women have for birth control. Thus, from a liberal feminist point of view, Seasonale and other menstrual-suppressing birth control pills would likely be embraced as expanding women's choices. Dr. Ruth Murkatz of the Population Council, who believes menstrual-suppression birth control pills provide "another choice for women, so they can control their destiny" and concludes "choice is good" (Chesler 2006), provides an example of this approach.

In addition, since liberal feminists tend to focus on equity, to the extent that menstruation makes it more difficult for women to compete equitably at work or to enjoy vacations and sex as much or as frequently as men, one could argue that, from a liberal feminist perspective, eliminating menstruation would be beneficial.

Radical Feminism

Whereas liberal feminism embraces the notion of expanded "choices" as a benefit in and of itself, radical feminism highlights how the choices offered to women are shaped by patriarchal systems and may in fact harm women. Radical feminists would likely deem Seasonale an antifeminist technology. Rather than suppressing menstruation, a radical feminist approach might be to redesign workplaces, schedules, and expectations to accommodate women's cyclically changing capacities and predilections. This is the position taken in Martin's classic *The Woman in the Body* (1987, 122–25), in which, after discussing the problems for women caused by late industrial society's demand for regimented physical and mental discipline while on the job, she refers to Beng women from the Ivory Coast as an example of a culture that plans on and accommodates a cyclic change in women's usual activities.

Socialist Feminism

This approach is based on the belief that "there is a direct link between class structure and the oppression of women" and that we must therefore challenge both "the ideologies of capitalism and patriarchy." Women must work side by side with men in order to achieve this (Stewart 2003). To the extent that menstrual-suppressing technologies are perceived as tools to render women more willing and able to be subjected to the physical and mental discipline that capitalism requires to maximize productivity and efficiency in the workplace (Martin 1987, 122), socialist feminists are likely to oppose them. Like the contributor to the MUM Web site who perceived a link between menstrual suppression and the way men and women are being "reshaped into suffering worker drones for capitalism," socialist feminists would likely prefer to organize labor in such a way that both men and women would be able to take more paid personal days.

Essentialist Feminism/Cultural Feminism

Essentialist feminism (now known more commonly as cultural feminism)[39] is based on the idea that "there are fundamental, biological differences between men and women, and that women should celebrate these differences.... Cultural feminists are usually non-political, instead focusing on individual change and influencing or transforming society through this individual change. They usually advocate separate female counter-cultures as a way to change society but not completely disconnect" (Stewart 2003). An example of this approach is found in a contribution to the MUM Web site from a woman in Chicago who writes, "I love my period. It's my Moontime, my time to relax, pamper myself, and be creative. ... I listen to female musicians, read female authors, admire female artists, and chat about intimate issues with my female friends. I eat healthier and indulge in the richest, darkest chocolate. I also feel a greater spiritual connection during my Moontime." Through this lens, menstrual-suppressing drugs are not only not feminist, but antifeminist.

Ecofeminism

This approach is premised on a deep link between women and nature, and patriarchy is understood as the simultaneous domination of both nature and women. In ecofeminism, women's special understanding of nature enables them to provide progressive solutions for "how humans can live in harmony with each other and with nature" (Stewart 2003). We saw examples of this in the contributions of those who feel their periods provide a special link to nature. These contributors tended to also embrace an essentialist/cultural perspective. Thus, it seems that from an ecofeminist point of view, like that of the essentialists/culturalists, menstrual suppression is not only not feminist, but antifeminist. However, because with drugs like Seasonale women will no longer be purchasing disposable, one-use only sanitary pads and tampons that end up in landfills (Strasser 1999, 161–70), one might argue that it therefore qualifies as an ecofeminist technology. The counterargument is that there are other technologies that already exist (e.g., reusable cotton pads, menstrual cups)[40] that meet this need without women having to suppress their cycles by making themselves dependent on a consumer product that must continuously be replenished.

African American Feminism

African American feminism focuses on "the promotion of black female empowerment" and is characterized by "the presentation of an alternative social construct for now and the future based on African American women's lived experiences." It is based on the recognition of "multiple systemic forces of oppression" and thus entails "fighting against race and gender inequality" (Barnes 2008, 1).

This perspective has been noticeably absent both in the public debate and in the

Lunapad "I love my period" round sticker. Permission and high-resolution images granted and sent by makers.

"Ditch Disposables" round sticker. Permission and high-resolution image granted by makers.

self-reports on the MUM Web site. (Only one of the contributors self-identifies as black and she is against suppression: "I am a black student doctor of a natural health care approach and a woman. I wouldn't stop menstruating if I had the chance (including after having children)." An African American feminist perspective would be cognizant of the many reproductive rights abuses that have been directed at African American women (as well as other women of color in the United States and globally). The early birth control movement was associated with the eugenics movement, which often singled out black women (Roberts 1997, 70–79). Forced sterilization and the targeting of blacks for new, inadequately tested, extended-regimen hormonal birth control in the form of Norplant and Depo Provera (Nelson 2003; Roberts 1997), suggests that African American feminism would be wary of Seasonale and other "extended-regimen" birth control pills.[41]

Existential Feminism

Existential feminism derives from the work of Simone de Beauvoir. In *The Second Sex* Beauvoir describes the ways women's bodies make them subservient to the demands of the species to procreate in a way that greatly exceeds the demands placed on men. She has a very negative view of menstruation, which she sees as a useless burden "from the point of view of the individual" (1989/1949, 27). She writes: "Menstruation is painful: headaches, over fatigue, abdominal pains, make normal activities distressing or impossible; psychic difficulties often appear: nervous and irritable, a woman may be temporarily in a state of semi-lunacy. . . . The body seem[s] a screen interposed between the woman and the world . . . stifling her and cutting her off" (1989/1949, 329).[42] She sees the end to menstruation with menopause as the only way that "women escape the iron grasp of the species" (1989/1949, 31).

From this, it seems evident that she would have embraced menstrual-suppressing drugs had they been available. This view is also supported in her general stance vis-à-vis nature. She celebrates human society that exerts mastery over nature. "Human society is an antiphysis—in a sense it is against nature; it does not passively submit to the presence of nature but rather takes over the control of nature on its own behalf" (1989/1949, 53). She has a generally positive attitude toward technology because of the potential to equalize men and women's physical abilities.[43]

However, other elements of *The Second Sex* suggest that she had some positive attitudes toward women's reproductive biology, linking ovarian functions to women's vitality (1989/1949, 27). But although she recognized the body as "one of the essential elements in her situation in the world," Beauvoir asserted "that body is not enough to define her as woman" (1989/1949, 87). Instead, Beauvoir privileges individual consciousness manifested through activities. Women's second-class status is a result of broader social traditions. She uses pregnancy as an example to illustrate the way the same biologic phenomenon differs depending on social arrangements. The burdens of maternity in societies where women do not have reproductive freedom and little social support "are crushing," whereas in societies where "she procreates voluntarily" and "society comes to her aid . . . the burdens of maternity are light and can be easily offset by suitable adjustments in working conditions" (1989/1949, 54). Thus, we might infer from this that she might favor similar social accommodations rather than drugs to alter women's biology.

Finally, Beauvoir urged women to "confront internalized desires that lead to acceptance, and thus perpetuation, of society's conventional definitions and expectations of femininity" (Marso 2006, 17). Hence, existential feminists might object to menstrual suppression as an unthinking compliance with conventional definitions of femininity; for example, the ability to wear a "pretty dress" without worry of staining. Furthermore, given Beauvoir's emphasis on individual intellectual and creative actualization, one would expect her to celebrate the achievements of women like this contributor to the MUM Web site who reports enjoying her greatest productivity during her menses. "I am an artist and a feminist article writer and I come up with my most powerful, eloquent, meaningful pieces during my Moontime."

Cyborg Feminism

According to Donna Haraway, the person most closely associated with cyborg feminism, the separation of nature and culture has been particularly detrimental to women (Haraway 1991). Cyborg feminists believe that disrupting the nature/culture duality will not only free us from the constraints of essentialism but will also allow us to conceive of and employ science and technology to further women's aims in a more nuanced manner. The cyborg—a heterogenous mix of human and machine—moves beyond binaries and essentialism. Among cyborgs, it becomes more difficult to distinguish dualistic categories such as "nature" and "culture." A cyborg feminist position poses unsettling (yet also liberating) questions: what counts as "nature"? What counts as a "woman"? The ambiguity, in Haraway's view, is liberating. Hence, a technology like Seasonale, which compels us to confront our cultural ideas of "normalcy," "nature," and "necessity," can be considered feminist for this reason. Haraway's cyborg is a compelling image because it reveals the arbitrariness, and thus instability, of our categorizations.

TWO SOCIALLY/MATERIALLY
SITUATED CONCLUSIONS

Given that whether or not a person considers menstrual suppressing birth control pills to be a feminist technology appears to be shaped both by one's personal experience with menstruation and the type of feminism(s) she or he embraces, it is not surprising that we (Layne and Aengst) differ in our assessment of this issue. We provide two alternative sets of conclusions as well as points of convergence.

Layne: I am a perimenopausal woman who has experienced regular, unremarkable menses for thirty-seven years or so. For the vast majority of my adult life I have used a diaphragm and contraceptive jelly for birth control, a method that I have found effective, safe, and easy to use, and one that has the added benefit of containing menses. I am pleased there are a growing number of ways to alleviate the suffering of women with "menstrual disorders." Given that two of the most of common therapies (hysterectomy and uterine ablation) preclude future child bearing, alternatives are clearly needed. But even for women with menstrual disorders, I am reluctant to deem menstrual-suppressing drugs as feminist technologies because the recent invention of the Vipon, a vibrating tampon that appears to relieve the menstrual pain of women with endometriosis (Vostral, this volume), alerts us to the fact that even if menstrual-suppressing birth control pills are the best option for women with menstrual disorders now, there is no reason to think that there are not better alternatives that have not yet been created by feminist designers.[44]

As for the use of menstrual-suppressing drugs to suppress "normal" menses (being cognizant, of course, of how both the notions of "normal/abnormal" are culturally shaped), as a radical feminist, I am opposed. Rather than women using technologies to alter themselves to more comfortably fit the demands of a patriarchally shaped world, I would prefer to reshape the world to better accommodate women. Furthermore, as a medical anthropologist working in the era of "big pharma" (Angell 2005), I am oriented against the expansion of drug regimens, and prefer mechanical (mostly reusable) menstrual-management technologies like menstrual cups (see illustration), diaphragms, reusable cloth pads (Vostral, this volume), or even menstrual extraction. Menstrual extraction is a technique developed by a group of feminists in Los Angeles at the Self-Help Clinic of the Feminist Women's Health Center, in which a woman or a friend inserts a thin tube into her uterus when her period begins, and uses a syringe to suction out the endometrial lining, a procedure that takes only minutes (Delaney, Lupton, and Toth 1976, 215–16). Delaney, Lupton, and Toth deemed menstrual extraction as "by far the most exciting discovery of the women's health movement" because it provides "newfound control over our own bodies" (1976, 217). This is a safer, less expensive, and more self-controlled alternative than daily drug use for those who wish to be spared the task of managing their menstrual cycles more conventionally. This technique has the added benefit of enabling early abortion, should that be desired.

A final consideration for me of the feminist/antifeminist valence of menstrual-suppressing drugs stems from the fact that these pills have been so newsworthy. Part of the reason these drugs make such good news copy is that women are so sharply divided on the subject. As we have seen from the contributions to the MUM Web site, not only do women differ physically (in their experience of menstruation) and attitudinally about the value/meaning of menstruation, they also are sometimes highly judgmental of those who differ from them. Could it be

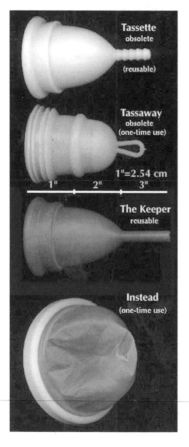

Menstrual cups.
Permission granted by
Henry Finley, Museum
of Menstruation.

that this subject is so mediagenic precisely because of its catfight potential? The tampon, the home pregnancy test, and the breast pump, technologies examined in this book, have not generated such heated divisions between women, nor have they been the subject of comparable media attention. I suggest that the valence of a technology, either to pit women against each other or bring them into solidarity, might be added to the list of criteria that qualify/disqualify technologies as feminist. Note, I am not suggesting that there should be a single feminism or that women could or should see things the same way. The diversity of the feminist movement is what makes it such "a many splendor'd thing" (Meyer 1987, 389) and, as noted in the introduction to this volume, this diversity will be an important resource for the proliferation of feminist technologies.

Aengst: I first became interested in Seasonale when my gynecologist told me that she could make my periods disappear. "Women don't have to have a period anymore," she said to me, "nowadays, there are just so many more options." I left the office disturbed by her comments, which seemed to be a bit too cavalier, although I was equally intrigued by the thought of having more reproductive health options. I realized that my own misgivings about taking long-term birth control pills was bound up in the fear of a new technology, essentialist ideas about gender, and fairly ingrained ideas about what was "natural." After much consideration, I am still uncomfortable with Seasonale and have no desire to use menstrual-suppressing drugs. Because I have been a regular user of "traditional" birth control pills for many years, Seasonale is appealing in its similarity to the pill regimen I am already taking. Yet, as someone who spent much of my adolescence wishing my periods were more regular, I find the idea of going many months without a period disturbing. Despite being aware of how the notion of "natural" is socially and culturally constructed, I find myself still preferring to have a monthly, more regular period.

Although I like that Seasonale disrupts notions of biological female essentialism, it leaves essentialist ideas about men uncontested.[45] In addition, this technology maintains the well-entrenched belief that women are the ones ultimately responsible for birth control. The development, distribution, and use of male methods of birth control is long overdue. Male-oriented birth control methods will not only disrupt gendered notions of female and male essentialism but will also challenge the idea that contraception is solely a "women's issue."

Furthermore, there are underlying class and sexist and racist ideologies related to menstrual suppression. Seasonale is an expensive technology that remains inaccessible to lower-income women and to those within the developing world. Many theorists have pointed out how cultural norms—such as determinations of who takes birth control pills and what methods are deemed "appropriate" for women in the developing world—influence those in the policy and development world, which ultimately determines where reproductive technologies travel (Sen, Germain, and Chen 1994). Unfortunately, because the development and policy literature often suggests that women in the developing world cannot be trusted to reliably take a daily pill, and because of the expense, menstrual-suppressing birth control pills are less likely to be available for women in the developing world.

Seasonale might very well be a useful technology for middle- and upper-class women who seek convenience and can afford to choose among many contraceptive technologies. Disrupting deeply entrenched norms of "nature" and "necessity" is a great step—and this is what makes Seasonale a worthwhile technological development—but it has not gone far enough.

CONCLUSIONS

Seasonale proves to be an excellent test case for honing the definition of feminist technology and for modeling a feminist technology assessment. As we have seen, menstrual suppression raises to the forefront differences among women—physical, social, cultural, and attitudinal. It also highlights differences within feminism. It is by struggling to take into account these differences that we can make headway in defining, recognizing, calling for, and creating feminist technologies.

NOTES

Thanks to Shirley Gorenstein and Deborah Johnson for helpful editorial suggestions; to Harry Finley, the director of the Museum of Menstruation for providing such a wealth of cultural data; to Si Ming Lee for tallying the MUM results; to Nancy Campbell, Ben Barker-Benfield, Ron Eglash, and Lori Marso for sharing their books and knowledge of feminist theory; and to Michael Halloran and Maral Erol for the helpful perspective their own work provided. Versions of this paper were presented at the Society for Social Studies of Science conference in Pasadena, to the Women's Studies Program at Union College, and at Society for Medical Anthropology/Society for Applied Anthropology meetings in Memphis, the Department of Anthropology and Women's Studies, State University of New York–Oneonta; Gender and Women's Studies, University of Illinois at Urbana-Champaign; and the Cambridge Interdisciplinary Reproduction Forum (CIRF) in Cambridge, England. Thank you to Sallie Hahn, Gail Landsman, Charlotte Faircloth, Sharra Vostral for those invitations. The essay was strengthened as a result.

1. Seasonale is a trademark of Duramed Pharmaceuticals, Inc.
2. Their actual effectiveness is estimated to be only 92 percent, with 8 of 100 women

on the pill getting pregnant each year. According to Gawande (2007), "[W]ith lower dose hormone formulations," like Seasonale, "missing a dose by even six hours puts a woman at" such risk he advises the use of condoms for that whole month.

3. These and other risks are discussed in detail in the six-page, small-print, black-and-white product insert.

4. This marketing slogan is trademarked by Barr Laboratories, of which Duramed is a subsidiary.

5. This is a classic case of the proliferation of what are known as "me-too" drugs, new products that "are no better than drugs already on the market to treat the same condition" (Angell 2005, 75). According to Angell (2005, 75), 77 percent of all new drugs approved by the FDA between 1998 and 2002 were "me-too" drugs. According to their Web site, www.shortperiod.com, "Loestrin 24 Fe uses the lowest dose of estrogen (20 mcg per pill) currently approved by the FDA for effective birth control."

6. Other hormonal contraceptive methods can also suppress menstruation. The contraceptive cervical ring, NuvaRing, is being used by some to suppress menstruation (Associated Press 2006a), and Implanon, a birth control rod implanted into the upper arm that works for three years "stops menstruation in many women." A feminist comedic video on menstrual-suppressing forms of birth control can be seen at http://www.feministing.com/archives/010078.html.

7. Respondents are encouraged to tell their age and where they are from.

8. Aengst asked sixty women on her e-mail list (who are white, middle-class, educated women, ages twenty-five to forty) whether they have ever taken birth control pills, whether or not they would take Seasonale, and why. Of the nineteen women who responded, eighteen had taken birth control pills; however, only four said they would take Seasonale. The two main reasons they gave for this was because they liked the monthly reassurance that they were not pregnant and they felt that taking Seasonale would disrupt the "natural" monthly cycle. In contrast, a much higher proportion of the 919 respondents to the question posed on the MUM Web site said they would suppress: 545 said they would, 374 said they would not. Women who post on the MUM Web site may be more likely to experience menstruation as problematic and have discovered the Web site while searching the web for help and support.

9. These physical differences include endometriosis, blood disorders, and mental (Thomas and Ellertson 2000, 922) or physical (Colligan 1994) disabilities that make managing menstruation particularly difficult.

10. See for comparison Hardon's discussion of the role of the Population Council in the development and testing of two other long-acting contraceptives, Norplant and antifertility vaccines, and the women's health movement's response (this volume).

11. Coutinho reports that his work with Depo Provera showed that "women who suffered from premenstrual tension and other menstrual disorders welcomed the long menstruation-free intervals." He moves directly from these particular women to generalize about "women"—"it was clear, that, contrary to conventional wisdom, women not only accepted the idea of not menstruating, they appreciated it as a benefit of the treatment"(1999, 10). Later, however, he reports on a ten-country study of menstruation

conducted by the World Health Organization in 1983, which found that the majority of respondents of all cultures related some physical discomfort and some mood changes linked to menstruation. Yet in what he sees as a "paradox," given "the many negative aspects of menstruation, . . . the majority . . . did not wish to use a contraceptive method that would suppress menstruation" (1999, 12).

12. They acknowledge their debt to Coutinho and Segal for their "ideas and suggestions imparted over many years" (Thomas and Ellertson 2000, 924).

13. Thomas and Ellertson (2000, 922) also draw on feminist rhetoric in questioning why "no other disease or condition that affects so many people on such a regular basis with consequences, at both the individual and societal level, is not prioritized in some way by health professionals or policy makers."

14. See Solinger (2001) on how "the language of choice" has come to replace "the language of rights" and how decisions about women's reproduction came to be cast in terms of "the individualistic, marketplace term 'choice.'"

15. According to Johnston-Robledo et al. 2006, 359), the first opponents (i.e., those whose views appeared in the popular press between 2000 and 2003) focused on safety issues rather than sociocultural or psychological concerns.

16. In 2006, Seasonale's sales reached $120 million (for the twelve months ending in June). A generic equivalent produced by Watson has since entered the market. Seasonale and Seasonique make up only 0.9 percent of the $1.7 billion annual U.S. market for oral contraceptives (Saul 2007, C4).

17. It is worth noting that there was comparable concern about the adoption by teens of tampons as a new menstrual-managing technology when they were introduced (Vostral, this volume).

18. In Turkey, when patients resist using hormone-replacement therapy (HRT) for menopause because they see HRT as "unnatural," some doctors argue "not everything natural is a good thing," such as floods and earthquakes (Erol 2008, 134).

19. This is a skillful and unconventional use of an enthymeme to persuade readers that it goes without saying that menstrual suppression is the preferred choice, but that we should be tolerant of those who make other, less enlightened, personal choices like choosing to continue menstruating.

20. It is worth noting that not all "natural," "real" periods involve ovulation. According to Weideger (1977, 6), "The majority of adolescents and the majority of women approaching menopause have cycles in which there is no ovulation, while most women in the 20–40 age group have menstruation without ovulation only once or twice a year."

21. An interesting comparison might be made with HRT for menopause. In Turkey, doctors respond to women who are reluctant to take HRT because hormones are "artificial" with a number of strategies, including arguing that "the estrogen that a menopausal woman takes [is a] part of nature. Like an apple tree presents the substance it takes from the earth to us as an apple and an apple is part of nature; the drugs that people make in the factories by substances they take from nature are the fruits of humans, so a piece of nature" (Kadayifci 2006, 37, quoted in Erol 2008, 134).

22. The Life Extension Institute of Palm Springs also offers anti-aging individualized

regimens of "Total Hormone Replacement Therapy" that "may include injections of testosterone and human growth hormone, topically applied testosterone gel, tablets of me-
latonin and as many as six other hormones that are supposed to slow the aging process and
intensify the patient's sense of well-being and sexual vigor" (Hoberman 2005, 13).

23. Again, HRT provides a fruitful comparison. According to radical feminists Germaine
Greer and Sandra Coney, HRT is an attempt by patriarchal medicine to "keep women
young and 'contributing,' if not to the continuation of the species, at least to the pleasure
of men (both sexually and temperamentally" (Roberts 2002, 39).

24. They describe religious prohibitions in the Koran and the Old Testament, and cite
a 1973 study by Karen Page that found the prohibition much more frequently observed by
Catholic and Jewish women than by Protestants. They also cite a study of black, medically
indigent women in Georgia among whom the taboo was "overwhelmingly observed" (1976,
22). In addition to religion, other explanations for the taboo include beliefs that it is bad
for men's health, for women's health, for the health of the unborn, and that women are not
aroused during their periods. They also mention the case of a woman who had an elective
hysterectomy "so that she [wouldn't] have to say no to her husband at *that* time" (1976, 23).
According to Coutinho and Segal (1999, 12), the majority of women respondents in the
WHO 1983 survey from all ten of the countries in the study (Egypt, India, Indonesia, Jamaica,
Mexico, Pakistan, the Philippines, Korea, United Kingdom, and Yugoslavia) "believed that
sexual intercourse should be avoided during menstruation."

25. Sometimes these prohibitions are religious. For example, according to Jewish law,
a man may not have sexual relations with his wife during menstruation nor for the seven
days following her bleeding, and even then not until she has performed the ritual purify-
ing bath, a mikvah (Alexander 2003). Menstruation is also considered ritually polluting
in Islam.

26. Several studies have found that men report more negative attitudes toward menstruation than women (Johnston-Robledo et al. 2006, 354), and one author suggests that
"as members of a culture that sexualizes or objectifies their bodies, [women] are motivated
to distance themselves or dissociate from bodily functions such as menstruation that
are deemed incompatible with their sexual attractiveness or desirability." Women who
reported higher levels of self-objectification had more negative attitudes toward menstruation (Johnston-Robledo et al. 2006, 354).

27. England reached this rate, one out of every five pregnancies, in 1997 (Edozien 1999).

28. About 10–15 percent are emergency c-sections.

29. The rate for caesarian births for first pregnancies increased to 29.2 percent, an
increase of more than 40 percent since 1996 (Bakalar 2005). A similar trend is seen in
Australia where the c-section rate has risen from about 5 percent in the 1970s to 19 percent in 1994, 27 percent in 2002, and 28.5 percent in 2003 (Hamer 2007a). Of these it is
estimated that 5–10 percent are scheduled at a woman's request (Hamer 2007b, 11).

30. Interestingly, Brazil, where Seasonale was developed, has particularly high rates,
with some hospitals reporting 80 percent of babies delivered this way (Park 2008).

31. It is not just users who are invited to assert their modernity by choosing this drug,
but physicians too. Coutinho and Segal (1999, 163) castigate those who subscribe to "the

traditional paradigm, ordained by Hippocrates in an era of medical naiveté, that regular menstruation is good for women." Thomas and Ellertson (2000, 922) link the belief that monthly menstruation is healthy to the outdated and "universally harmful medical practice" of therapeutic, induced "bleeding" of "previous centuries." Similar arguments are made by Turkish doctors in the face of resistance to HRT. The doctor of one woman who explained that she wanted to stop HRT because "her grandmother or her mother never took anything and they were fine" replied that they also rode in ox-carts instead of taking the plane (Erol 2008, 134).

32. http://www.clevelandclinic.org/health/health-info/docs/3200/3296.asp?index=11283.

33. The woman who goes on family vacations, attends yoga classes, travels for business, and can choose fun, distinctive shoes from ample consumer choices is the woman who can afford Seasonale.

34. Of the three types of appropriation delineated by Eglash (2004:x-xii), this represents an example of "adaptation," which involves a change in use but not structure, and also illustrates the collective force of consumers in shaping technology design through marked demands" (Eglash 2004:xvi).

35. noperiod.com and Cox, Amy and Christy Feig (September 8, 2003) "New Birth Control to Limit Women's Periods" CNN.

36. Rosser (2006) provided a model for this section.

37. The same contributor writes, "I never would have thought that in the 21st-century, women would feel this way about their own body!" She makes a distinction for women who have very painful periods, and those who complain that "it smells" or that it's "disgusting," ". . . having my period doesn't make me dirty or repulsive. It's not disgusting. Do you say blood is disgusting when you cut yourself? I don't think so. You may even automatically lick it when it's a little scratch or something like that (please don't shoot me! I'm not telling we should do the same with the period). But period should be disgusting because it comes out of your sex? This way of thinking . . . shows how much women don't really love or accept themselves. They consider their body beautiful as long as it's attractive to the opposite sex: how nice it is to have big tits nowadays (even if it means back pains or problems, even if it has to be achieved through surgery and looks completely fake and unnatural)! Guys love it. But how disgusting it is having your period: it's not attractive to men . . . "Why can't we women be proud of what we are, no matter if it is pleasant to men or not."

38. According to Georges (2009:100) in Greece menstrual blood and semen are considered similar but in both cases, the discharge of these bodily fluids is understood to rid men's and women's bodies of accumulated impurities (of male and female dirt) and the regular expulsion of both are considered essential for health.

39. Probably as the result of what Fuss (1989:1) describes as "paranoia around the perceived threat of essentialism."

40. The Lunapads website argues, "Like recycling bottles and newspapers, washing Lunapads or rinsing out the DivaCup is a little more work than throwing away your used pads and tampons. But with over 14 billion pads, tampons and applicators going into North American landfills every year, it's a small but important way of taking personal responsibility for a massive environmental problem." www.lunapads.com.

41. The same may be true for other methods of menstrual suppression such as hysterectomies. Whereas black women have too often been urged or coerced into having hysterectomies, contributors to the MUM Web site, who are presumed white unless they mention their race or ethnicity, complain about their difficulties in obtaining surgical menstrual-suppression. For instance, a self-reportedly healthy woman who has never had bad cramps just heavy bleeding writes, "I'm interested in other forms of suppression since they refuse to give me an elective hysterectomy. Nor will they offer me endometrial ablation . . . would love something permanent . . . I even asked the vet if I could be spayed along with the cat. He just laughed. He thought I was kidding." Similarly, a 43-year-old reports "from home due to missing yet another day from work because of my periods," of her inability to get elective surgical suppression. "I had my tubes tied 11 years ago, and the doctor at that time refused to do a hysterectomy or oophorectomy [removal of the ovaries] to stop my periods, saying I was "too young." Another woman, a stripper who laments the trouble her periods cause her at work has been denied a hysterectomy. "I have to work very hard to conceal my period. The club where I work will not give you time off for your period so here I am trying to find ways to conceal my period while dancing nearly nude . . . most gyno's won't even consider giving me a hysterectomy since "I have nothing wrong." I tried the Norplant, Depo, and now the Seasonale pill. I still have my period on all those things." One woman who did get a hysterectomy explains how happy she was to do so, "I did [stop menstruating]! I have had horrible periods for years, so much so that I missed many professional and personal obligations because of them and they became near-constant and incapacitating. Happily, last week, at age 37 I had a hysterectomy. No qualms about it really and glad to be done with the whole thing."

Others report having used Depo or Norplant, both of which suppress menstruation. For example, a 40 year old who looks forward to menopause reports how much she "enjoy[ed] the year and a half that I was using Depo Provera for birth control. I didn't have a period for nearly two years. It was AWESOME!!!!!!" Others report having tried them but needing to stop because of side effects.

42. She reports that "almost all women—more than eighty-five percent—show more or less distressing symptoms during the menstrual period" (1989:28–29).

43. This strand of her thinking was taken up and developed by Shulamith Firestone, one of Beauvoir's most well-known heirs, in *The Dialectic of Sex*, which she dedicated to Beauvoir. Firestone (1972:8) writes, women "throughout history before the advent of birth control were at the continual mercy of their biology—menstruation, menopause, and 'female ills,' . . . all of which made them dependent on males." In her view, "it was nature, then, not history, that underlay the inequality between the sexes" (Meyer 1987:396). Firestone (1972:10) asserted, "the 'natural' is not necessarily a 'human' value. Humanity has begun to outgrow nature." Technology, she believed, provided the means for women's liberation from their biology. Hence, menstrual suppressing drugs that help women "outgrow nature" and "liberate them from their biology," would thus be supported by this strand of existential feminism.

44. As always, technologies that would address cause rather than symptoms would be preferable.

45. According to Oudshoorn, "only about 17 percent of contraceptive users rely on so-called male methods."

References

Alexander, Elizabeth Shanks. 2003. "Healing Waters: Turning to Jewish Ritual After Miscarriage." *Moment* (October): 50–55, 70–72.

Angell, Marcia. 2005. *The Truth about the Drug Companies: How They Deceive Us and What to Do about It.* New York: Random House.

Associated Press. 2006a. "Pills Rendering Menstrual Period Optional" (May 22). http://www.intelihealth.com/IH/ihtIH/WSIHW000/333/22002/466295.html?d=dmtICNNews.

———. 2006b. "New Birth Control Products Block Periods" (May 22).

http://www.intelihealth.com/IH/ihtIH/WSIHW000/333/22002/466296.html?d=dmtICNNews.

Bakalar, Nicholas. 2005. "Premature Births Increase along with C-Sections." *New York Times* (December 22).

———. 2007. "Optional Caesareans Carry Higher Risks, Study Finds." *New York Times* (March 27).

Barnes, Shellie. 2008. *African American Feminisms: A Multidisciplinary Bibliography.* Updated April 16. University of California at Santa Barbara Libraries. http://www.library.ucsb.edu/subjects/blackfeminism/introduction.html.

Beauvoir, Simone de. 1989 [1949]. *The Second Sex.* Translated and edited by H. M. Parshley. New York: Vintage.

Brody, Jane E. 2003. "With Childbirth, Now It's What the Mother Orders." *New York Times* (December 9).

Bucek, Amelia. 2005. "Beyond the Hype: What You Should Know about the Seasonale Birth Control Pill." *Different Takes* 36(Spring) http://popdev.hampshire.edu/projects/dt/dt36.php.

Chesler, Giovanna, Director/Producer. 2006. *Period: The End of Menstruation?* The Cinema Guild.

Colligan, Sumi Elaine. 1994. "The Ethnographer's Body as Text: When Disability Becomes 'Other'—Abling." *Anthropology of Work Review* 15(2&3): 5–9.

Coutinho, Elsimar, and Sheldon J. Segal. 1999. *Is Menstruation Obsolete?* New York: Oxford University Press.

Cox, Amy, and Christy Feig. 2003. "New Birth Control to Limit Women's Periods." CNN (September 8).

Davis, Jean Lerche. 2003. "FDA Approves New Birth Control Pill Seasonale—For Seasonal Periods. *Free Republic* (September 5). http://www.freerepublic.com/focus/fr/979259/posts.

Delaney, Janice, Mary Jane Lupton, and Emily Toth. 1976. *The Curse: A Cultural History of Menstruation.* New York: E. P. Dutton.

Dixon-Mueller, R. 1993. *Population Policy and Women's Rights: Transforming Reproductive Choice.* Westport, Conn.: Praeger.

Duramed. 2003. Seasonale (levonorgestrel/ethinyl estradiol tablets) 0.15 mg/0.03 mg Rx only. Product insert. Pomona, N.Y.: Duramed Pharmaceuticals.

——. 2005a. Thank You for Asking about Seasonale. Form letter. Pomona, N.Y.: Duramed Pharmaceuticals.

——, 2005b. Extended-Regimen Seasonale: The Daily Birth Control Pill That's the Same but Different. Brochure. Pomona, N.Y.: Duramed Pharmaceuticals.

Edozien, L. C. 1999. "What Do Maternity Statistics Tell Us about Induction of Labour?" *Journal of Obstetrics and Gynaecology* 19, no. 4: 343–44.

Eglash, Ron. 2004. "Appropriating Technology: An Introduction." In Ron Eglash, Jennifer L. Croissant, Giovanna Di Chiro, and Rayvon Fouche, eds., *Appropriating Technology: Vernacular Science and Social Power,* vii–xxi. Minneapolis: University of Minnesota Press.

Erol, Maral. 2008. Rites of the Second Spring: Situational Analysis of Postmenopausal Hormone Replacement Therapy in Turkey. Unpublished dissertation, Science and Technology Studies, RPI, Troy, N.Y.

Fathalla, Mahmoud. 1994. *Fertility Control Technology: A Woman-Centered Approach in Population Polices Reconsidered: Health, Empowerment, and Rights,* 223–34. Boston: Harvard School Public Health.

Fink, Jennifer L. W. 2000. *Labor Inductions on the Rise: A Product of Our Times.* American Baby.com. (Accessed May 2008.)

Firestone, Shulamith. 1972 [1970]. *The Dialectic of Sex: The Case for Feminist Revolution.* New York: Bantam.

Frank Fox, Mary, Deborah Johnson, and Sue Rosser, eds. 2006. *Women, Gender, and Technology.* Urbana: University of Illinois Press.

Fuss, Diana. 1989. *Essentially Speaking: Feminism, Nature and Difference.* New York: Routledge.

Gawande, Atul. 2007. "Let's Talk about Sex." *New York Times* (May 19): A13.

Georges, Eugenia. 2009. *Bodies of Knowledge: The Medicalization of Reproduction in Greece.* Nashville, Tenn.: Vanderbilt University Press.

Gladwell, Malcolm. 2000. "John Rock's Error: What the Co-Inventor of the Pill Didn't Know about Menstruation Can Endanger Women's Health." Annals of Medicine. *New Yorker* (March 10). http://www.gladwell.com/2000/2000_03_10_a_rock.htm.

Gordon, Linda. 2002. *The Moral Property of Women: A History of Birth Control Politics in America.* Urbana: University of Illinois Press.

Gottlieb, Alma. 2002. "Afterword." *Ethnology* 41, no. 4: 381–90.

Hamer, Michelle. 2007a. "The Great Caesar Debate." *Sydney Morning Herald* (May 10).

——. 2007b. *Delivery by Appointment: Caesarean Birth Today.* Sydney: New Holland.

Haraway, Donna. 1991. *Simians, Cyborgs, and Women: The Reinvention of Nature.* New York: Routledge Press.

Harding, Sandra. 1998. *Is Science Multicultural? Postcolonialisms, Feminisms, and Epistemologies.* Bloomington: Indiana University Press.

Harrell, Barbara B. 1981. "Lactation and Menstruation in Cultural Perspective." *American Anthropologist* 83, no. 4: 796–823.

Harris, Shayla, and Stephanie Saul. 2006. *Against the Flow.* Video. Business. *New York Times.* http://video.on.nytimes.com/index.jsp?fr_story=20bfd568e942f82f7acd590c3 39e10ecd69e6fd2. (Accessed April 20, 2007.)

Hoberman, John. 2005. *Testosterone Dreams: Rejuvenation, Aphrodisia, Doping.* Berkeley: University of California Press.

Hogle, Linda F. 2005. "Enhancement Technologies and the Body." *Annual Reviews in Anthropology* 34: 695–716.

Houck, Judith A. 2003. "'What Do These Women Want?': Feminist Responses to Feminine Forever, 1963–1980." *Bulletin of the History of Medicine* 77: 103–32.

Johnston-Robledo, Ingrid, Jessica Barnack, and Stephanie Wares. 2006. "'Kiss Your Period Good-Bye': Menstrual Suppression in the Popular Press." *Sex Roles* 54: 353–60.

Kelley, Tina. 2003. "New Pill Fuels Debate over Benefits of Fewer Periods." *New York Times* (October 14, 2003).

Lane, Brenda. 2006. "Labor Inductions on the Rise." *Suite101* (August 13). http://pregnancychildbirth.suite1010.com/article.cfm/labor_inductions_on_the_rise.

Layne, Linda L. 2003. *Motherhood Lost: A Feminist Account of Pregnancy Loss in America.* New York: Routledge.

Marso, Lorie Jo. 2006. *Feminist Thinkers and the Demands of Femininity: The Lives and Work of Intellectual Women.* New York: Routledge.

Martin, Emily. 1987. *The Woman in the Body: A Cultural Analysis of Reproduction.* Boston: Beacon Press.

Meyer, Donald. 1987. *Sex and Power: The Rise of Women in America, Russia, Sweden, and Italy.* Middletown, Conn.: Wesleyan University Press.

National Women's Health Network. 2004. Menstrual Suppression, Extended Cycle Oral Contraceptive Pills, and Seasonale. Fact sheet. (April).

Nelson, Jennifer. 2003. *Women of Color and the Reproductive Rights Movement.* New York: New York University Press.

O'Grady, Kathleen. 2001. "Are Periods Passé? A Review of *Is Menstruation Obsolete? How Suppressing Menstruation Can Help Women Who Suffer from Anemia, Endometriosis, or PMS.*" *Herizons Magazine* (Winter). www.herizons.ca/node/42.

Ortner, Sherry. 1974. "Is Female to Male as Nature Is to Culture?" In John McGee and Richard Worms, eds., *Anthropological Theory,* 402–13. Mountain View, Calif.: Mayfield.

Oudshoorn, Nelly. 1994. *Beyond the Natural Body: An Archeology of Sex Hormones.* London: Routledge.

———. 2003. *The Male Pill: A Biography of a Technology in the Making.* Durham, N.C.: Duke University Press.

Park, Alice. 2008. "Womb Service: Why More Women Are Making Caesarians Their Delivery Choice." *Time Magazine* (April 28): 65–66.

Prior, Jerilynn C., and Christine L. Hitchcock. 2006. Manipulating Menstruation with Hormonal Contraception—What Does the Science Say? 1–5. Centre for Menstrual Cycle and Ovulation Research. http://www.cemcor.ubc.ca/articles/misc/manipulation-menstruation.shtml. (Accessed May 20, 2008.)

Rako, Susan. 2003. *No More Periods? The Risks of Menstrual Suppression and Other Cutting-Edge Issues about Hormones and Women's Health.* New York: Harmony Books.

Roberts, Celia. 2002. "'Successful Aging' with Hormone Replacement Therapy: It May Be Sexist, but What if It Works?" *Science as Culture* 11, no. 1: 39–59.

Roberts, Dorothy. 1997. *Killing the Black Body: Race, Reproduction, and the Meaning of Liberty.* New York: Vintage.

Rosser, Sue V. 2006. "Using the Lenses of Feminist Theories to Focus on Women and Technology." In Mary Frank Fox, Deborah Johnson, and Sue Rosser, eds., *Women, Gender, and Technology,* 13–46. Urbana: University of Illinois Press.

Sanabria, Emilia. 2008. Limits That Do Not Foreclose: Biomedical Intervention, Hygiene and Sex Hormones in Salvador, Brazil. Ph.D. dissertation in Social Anthropology, King's College, Cambridge.

Saul, Stephanie. 2007. "Pill That Eliminates the Period Gets Mixed Reviews." *New York Times* (April 20): A1, C4.

Sen, Gita, Adrienne Germain, and Lincoln C. Chen. 1994. *Population Policies Reconsidered: Health, Empowerment and Rights.* Cambridge, Mass.: Harvard University Press.

Shaw, Gina. 2003. The No Period Pills, WebMD (September 8). http://my.webmd.com/content/article/71/81215.htm?z=1689_00001_2418_00_03.

Smith, Carol. 2003. "The Pill Indeed Can Stop Periods." *Seattle Post-Intelligencer* (April 3).

Society for Menstrual Cycle Research. 2003. Menstrual Suppression. http://menstruation-research.org/position/menstrual-supression/. (Accessed May 2007.)

Solinger, Rickie. 2001. *Beggars and Choosers: How the Politic of Choice Shapes Adoption, Abortion, and Welfare in the United States.* New York: Hill and Wang.

Stewart, Cara. 2003. Different Types of Feminist Theories. http://www.colostate.edu/Depts/Speech/rccs/theory84.htm#cultural. (Accessed May 2007.)

Strasser, Susan. 1999. *Waste and Want: A Social History of Trash.* New York: Henry Holt and Company.

Thomas, Sarah, and Ellertson, Charlotte. 2000. "Nuisance or Natural and Healthy: Should Monthly Menstruation be Optional for Women?" *The Lancet* 355, no. 9207 (March): 922–24.

Vostral, Sharra L. 2003. "Reproduction, Regulation, and Body Politics." *Journal of Women's History* 15, no. 2 (Summer): 197–207.

———. 2005. "Masking Menstruation: The Emergence of Menstrual Hygiene Products in the United States." In Andrew Shail and Gillian Howie, eds., *Menstruation: A Cultural History,* 243–58. New York: Palgrave.

Wajcman, Judy. 1991. *Feminism Confronts Technology.* University Park, Pa.: Penn State University Press.

———. 1994. *Delivered Into Men's Hands? The Social Construction of Reproductive Technology in Power and Decision: The Social Control of Reproduction.* 153–75. Cambridge, Mass.: Harvard School of Public Health.

Weideger, Paula. 1977. *Menstruation and Menopause* (revised and expanded). New York: Dell.

Web sites

http://www.feministing.com/archives/010078.html
www.noperiod.com
www.womenshealthnetwork.org

Blogs

http://www.electrolicious.com/archives/2003/09/seasonale.html
http://groovychk.livejournal.com/85697.html
http://thewelltimedperiod.blogspot.com/

3 Why the Home Pregnancy Test Isn't the Feminist Technology It's Cracked Up to Be and How to Make It Better

LINDA L. LAYNE

INTRODUCTION

HOME PREGNANCY TESTS (initially known as "do-it-yourself kits") are relatively low cost and easy for women in the United States to obtain and use.[1] In addition, they are noninvasive, pose no apparent health risks, and boast high levels of accuracy. In other words, they appear to be the very type of technology advocated by the women's health movement and by science and technology studies (STS) scholars who seek more democratic design and use of technoscience. Yet, examination of ninety-two first-person accounts of use posted between 2003 and 2005 as part of a U.S. National Institute of Health (NIH) project on the history of the test, accounts published on a pregnancy Web site, and the newsletters of two U.S. pregnancy loss support organizations spanning the period of 1981 to 2004, as well as an opportunistic sample of home pregnancy tests purchased in the United States, Canada, China, Argentina, and Uruguay, suggest that the presumed benefits of this technology are not so clear.[2] In fact, there are a number of hidden costs that come into relief when we examine how and by whom they are used. I conclude that home pregnancy tests do not offer women the benefits they purport to do and, in fact, in some ways disempower women by deskilling them, devaluing their self-knowledge, and enticing them to squander their buying power on frivolous consumer products. Despite all this, home pregnancy tests may be of value to some women in some circumstances, and thus, these information technologies should be improved to better serve women.

HISTORY OF HOME PREGNANCY TESTS

Sarah A. Leavitt (2005) provides a history of the home pregnancy test on the Web site "A Thin Blue Line" that she developed for NIH while a staff historian there. (Material on the history of the tests and any first-person accounts not otherwise acknowledged are from this site.) Leavitt begins the history of urine-based pregnancy testing with an apparently fairly (70 percent) reliable test used in ancient Egypt: a pregnant woman's urine would cause barley or wheat to grow; the urine of a woman who was not pregnant would not have this effect. Gelis (1991, 48) also reports urine tests for pregnancy in early modern Europe.

Hormones were named as such in the 1890s, the same decade, according to Leavitt (2005), that women were encouraged to seek prenatal care as soon as they realized they were pregnant. Progesterone was named in 1903 and isolated in 1934. During what is known as the "heroic age of reproductive endocrinology" (1926–40), "the chief naturally occurring estrogens, androgens, and progesterone were isolated and characterized, and the hypophyseal (anterior pituitary), placental, and endometrial gonadotrophins were also discovered" (Clarke 1998, 122). During this same period, "pregnancy diagnosis" was one of the "reproductive problems" addressed by researchers interested in improving the quality and quantity of livestock production (Clarke 1998, 45). As Oudshoorn (1994) observes, the study of sex hormones "focused almost exclusively on the female body." Gynecological practices made "the female body . . . an easily accessible supplier of research materials and convenient guinea pig for tests" (Oudshoorn 2003, 5).

A pregnancy test was devised in 1927 by Aschheim and Zondek (known as the A-Z test), which identified the presence of hCG (human chorionic gonadotropin) in urine. To test for pregnancy, a woman's urine was injected into an immature rat or mouse. If the subject was not pregnant, there would be no reaction. In the case of pregnancy, the rat would show an estrous reaction (be in heat) despite its immaturity. This test implied that during pregnancy there was an increased production of the hormone. During early studies of the A-Z test, the scientists discovered that testicular tumors could produce hCG as well. By the late 1930s, the simpler and faster Hogben test using frogs had become popular in Europe and North America (Gurdon and Hopwood 2000).[3] Because most of these bioassays were unable to distinguish between hCG and luteinizing hormone (LH) except at extraordinarily high rates of hCG, they were not reliable tests for early pregnancy.

In the 1960s, an immunoassay for pregnancy was developed by L. Wide and C. A. Gemzell but it produced many false negatives and positives and was not sensitive enough to diagnose early pregnancy. In 1966, A. R. Midgley described the first radioimmunoassay for hCG, but the test still could not differentiate between hCG and LH. Several other laboratories reported improvements on this test, but did not solve this basic problem.

Judith Vaitukaitis and Glenn Braunstein finally solved it after they joined the NIH in 1970. At that time it was known that the body secreted hCG during pregnancy and with certain types of cancer. It was clear that there would be significant benefits if accurate levels of hCG could be measured both in terms of cancer[4] and pregnancy. Dr. Vaitukaitis identified a subunit that was unique to hCG and then discovered an antibody specific to this subunit. Vaitukaitis, Braunstein, and Ross published a paper in 1972 describing the test that could finally distinguish between hCG and LH. They realized from the start the great commercial potential of this finding in terms of an early pregnancy test, and tried, without success, to get NIH to allow them to patent it. It is important to recall that at the time of this breakthrough, abortion was still illegal in the United States.

Home pregnancy tests became available in Western Europe and Canada before they did in the United States (Yankauer 1976). One woman who contributed her story of test use to the NIH Web site remembers driving from Detroit to Ontario, Canada, in December of 1971 to buy a kit for a college friend of hers who was "terrified that she might be pregnant and didn't want to go to the university health service to find out."

According to the 1976 edition of *Our Bodies, Ourselves*, published by the Boston Women's Health Book Collective (BWHBC), the year before home tests became available in the United States, there were at that time four ways of testing for early pregnancy: a urine test, a blood test, a hormone withdrawal test that was banned by the U.S. Federal Drug Administration (FDA)[5] in 1975 but apparently continued to be used by some doctors in 1976, and a pelvic exam. All of these could be done by "doctors, midwives, nurse-practitioners, physicians' assistants" at their offices or "at family-planning organizations, women's health centers and abortion clinics." The urine or blood tests could also be done at a laboratory listed in the yellow pages, although not all labs would give results directly to the woman. One could also do a urine test through the mail for eight dollars (Boston Women's Health Book Collective 1976, 221). The Boston Women's Health Book Collective goes on to explain that "verifying a pregnancy takes two procedures: a laboratory test which checks the urine for hCG . . . and a pelvic examination by a trained person to check for relevant changes in your cervix and uterus" (Boston Women's Health Book Collective 1976, 221).

In 1976 the FDA approved four home pregnancy tests that were deemed "substantially equivalent"[6] and by 1977 e.p.t. had reached the market. These products have multiplied over the years; in 2003 a Web site offered a comparison of fifty-two different brands available in the United Kingdom and/or United States (Fertility Plus 2003).

Over time, the tests have been improved, becoming faster and easier to use. The first e.p.t. test consisted of "a vial of purified water, a test tube containing, among other things, sheep red blood cells . . . as well as a medicine dropper and clear plastic support for the test tube, with an angled mirror at the bottom"; required the first urine sample of the day; and took about two hours. One woman recalls getting up one morning in 1978 to collect her urine, but since "it took two hours for the result to come in . . . I had to refrigerate the urine until I came home from work later that day. The test could not be disturbed. You had to put it where it would not feel any vibration." A woman who used the test in 1983 when she was sixteen and living at home with her parents who did not know that she was having sex, remembers having "to set it up in my cupboard very carefully so it wouldn't be discovered or knocked over." A woman who used a test in 1989 recalls, "It was not easy to use. . . . You pretty much felt like a chemist. . . . There were droppers to put drops of urine into a tube, you had to shake it up and then put this stick with little white beads in the end into the tube and wait something like 10 or 15 mins." Another woman who first used the tests in 1988 reports that back then, it was "not so easy to use. . . . The new tests are so much easier to take!"[7] Although some prefer the "pee on the stick" models and others prefer the cup, and some find the new digital models "cool" whereas others find them difficult to use, the consensus is that since their introduction, home pregnancy tests have "gotten easier and easier."[8]

Although the home (the woman's, her boyfriend's, or a friend's) appears to be the most common place the tests are used, "home pregnancy tests" are also used in other locations, including bathrooms at work, hotel rooms, dormitories, barracks, and bathrooms at the store where it is purchased.

USERS

Feminist scholars have been instrumental in expanding technology studies scholarship to include users. This turn to "the consumption junction" is attributed to Cowan's work in the late 1970s on domestic technologies. Since then, many STS scholars have examined how women and men actually use technologies. More recently, attention is also being given to how "presumed users" are "configured" during the design process and by journalists and people working in the public sector (Oudshoorn and Pinch 2005, 8–9, on Woolgar, Akrich, van Kammen, and Epstein). Another relatively new interest is in how users can act as "agents of technological change" by "appropriating" technologies and using them for their own purposes. Feminist scholars have also been particularly attentive to the diversity of users (Oudshoorn and Pinch 2005). In this essay I discuss the presumed and actual users of home pregnancy tests, drawing attention to differences among users, including those who have a pregnancy loss.

At first, the presumed users seem clear—the tests are designed for women.[9] But not all women, only those who are potentially pregnant; that is, those who in public health discourse are considered "at risk for pregnancy." This means they (1) must be of child-bearing age and (2) have been exposed to sperm. Even within the category of "possibly pregnant women" we find fundamental differences. In fact, there are two very different sets of presumed users: (1) women who wish to be pregnant, and (2) those who do not.[10] This fundamental difference is reflected in the fact that in *Our Bodies, Ourselves*, information on pregnancy testing is presented in two different locations: first in the chapter on abortion and then again in the chapter on pregnancy (Boston Women's Health Book Collective 1976). This raises interesting design issues because a single product is intended for users who have the opposite goals. What these two sets of users share is a perceived "need to know" and this points to the fact that home pregnancy tests are fundamentally information technologies.

Presumed and actual users are socially and physically diverse and include women from all walks of life, classes, ethnic backgrounds, and a wide range of ages, and, in the United States, are expected to speak either English or Spanish.[11] An important subgroup of presumed users are the estimated "6 million women who are challenged by infertility . . . [and are] trying actively to get pregnant." They are seen as a particularly promising market for home pregnancy tests because they "tend to be quite anxious to find out if they have been successful" (Johnsen 2003).

Of those who use the test, some are pregnant; some are not. Of those who are pregnant, regardless of whether they are in the "want to be pregnant" or "don't want to be pregnant" group, some will have viable pregnancies and others won't. In other words, some of those who are thrilled to learn they are pregnant will suffer a pregnancy loss (the rate is 15–20 percent with an additional 10 percent loss rate during the days between conception and first missed menses (Wilcox et al. 1988),[12] and of those who are dismayed to learn they are pregnant and undergo an abortion, the same portion (15–20 percent) would have lost the pregnancy naturally anyway.

Furthermore, women are not the only users. First-person and fictional accounts of pregnancy test use regularly feature men who go out to buy the tests, hover nearby while they are being used, are consulted for interpretation of the results, and are consulted—or not—on how to proceed given the results. These men are generally sexual partners or spouses but in one of the accounts posted on the NIH Web site, a father tells of going out to buy a third test because he was sure his daughter could not be pregnant.[13] More typical is Greg, who reports: "[W]e purchased" the test, "we followed the instructions," and "discover[ed] that we were pregnant. We were overjoyed." In Swain's (2004, 62–63) fictional account, the heroine's partner, Eddie, runs out to buy the test and returns with three tests, folic acid, and a fake rose, then stands outside the door shouting instructions about how to use it ("Only pee on it for five seconds. . . . Lay it flat. On a sink. . . . Don't hold it up."), and pestering her with questions, "Are you done yet?" "Everything go ok?"

THE PURPORTED BENEFITS OF THE HOME PREGNANCY TEST

By and large, feminists have embraced the home pregnancy test. The 1984 edition of *Our Bodies, Ourselves*, the first edition to come out after the advent of home tests in the United States, observed that some women "feel isolated doing a home test" but that others "appreciate the option of a home pregnancy test because it gives privacy, convenience and control over the experience" (Boston Women's Health Book Collective 1984, 285) and mention two advantages to "find[ing] out early. If you want to have a baby, you can take extra good care of yourself. If you decide not to continue the pregnancy, you can get an early abortion" (Boston Women's Health Book Collective 1984, 284). Another early proponent of these kits argued that do-it-yourself pregnancy tests would "help women gain the control over their bodies which is their right." She noted the importance of having these products "available over-the-counter so that we are not dependent on medical super-structures for confirmation of the outcomes of our own reproductive choices" (Oakley 1976, 502). A 1978 *Mademoiselle* article evaluating e.p.t. soon after it reached the U.S. market observed that a home test (1) spares women having to "wait several . . . weeks for a doctor's confirmation"; (2) offers more "privacy"; (3) "gives you a chance, if pregnant, to start taking care of yourself"; (4) "or to consider the possibility of early abortion" (Leavitt 2005). In the following sections I probe these purported benefits.

Providing Knowledge Directly to Women

At first glance, it appears that a home pregnancy test takes power/knowledge out of the hands of experts and places it in the hands of women. Notably, opposition to these kits came from professional laboratory technicians who saw the tests as undermining their authority.[14] However, despite the fact that these tests boast a very high accuracy level, accounts by users and representations of use in popular culture indicate that they are not considered authoritative by women or health care providers. Authoritative knowledge is "the knowledge that participants agree counts in a particular situation, that they see as consequential, on the basis of which they make decisions and provide justifications for courses of action. It is the knowledge that within a community is considered legitimate, [and] official" (Jordan 1993/1978, 152–54).

Women often do not trust the results of the test either because they believe the product may be flawed, or they fear they have erred in using it and so perform repeat tests. Most of the women who contributed accounts to Leavitt's NIH history project tell of having used multiple tests. One woman who takes "the Pill . . . religiously, but still worrie[s] [that she'll] be part of the 1 percent ineffective group," reports frequent use of the kits. "I took one yesterday—actually, I took two, since drinking lots of water can apparently mask a positive, and I'd had 2 glasses that morning, so I didn't trust the negative and took another test a few hours later to reassure myself." Another woman reports, "When I got the result, I didn't believe it. So I had him go to the store and get a different brand and I took it and it was positive, then I said, 'Take me to Planned Parenthood.' I had three different types of tests in one day and all came out positive so I had to realize that day that I was pregnant."

The most dramatic example of multiple test use occurred in 1991 on Murphy Brown, a popular American television comedy that gained notoriety when the vice president, Dan Quayle, criticized the character, a young, single, professional woman, for having a baby out of wedlock with no father in the picture. Over the course of two episodes, Murphy Brown took twenty tests, all of which were positive (Leavitt 2005).[15] More recently, the 2007 Oscar-winning movie Juno opens with another unwed woman using multiple tests, all of which are positive. The movie opens with the sixteen-year-old protagonist buying, using, and reading her third positive pregnancy test of the morning in her town's general store while engaging in witty banter with the male sales clerk and another female customer.

Women sometimes doubt their own perceptions and question their ability to accurately read the home test results. For example, a contributor to a pregnancy loss support newsletter recalls, "For the last eleven years, I'd been trying to conceive. Month after month, prayer after prayer, and one home pregnancy test after another. . . . At one point, I believe I psychologically made myself pregnant. I had all the typical symptoms, but I still got a minus sign." Then after she had finally decided to "accept the fact that I would never be a mother," Chrissy's "dream came true . . . I saw a plus sign! Hallelujah! My first thought was 'Thank God I don't have to buy anymore of those pregnancy tests' (I should have been buying stock in them instead). My second thought was 'Is it really a plus sign or are my eyes playing a trick on me?'" She woke her "significant other" so that "he could confirm the test result" (Coggins 2000).

Other such accounts are found on the Baby Corner Web site on which some women post their pregnancy journals. For example, the entry titled "The Nine Month Journey of Life" by Teresa-Lynn begins, "Is that a line? Does this look like a line? I think I see a line, do you see a line? Those were my first stuttered out shocked words as I stared at one of the home pregnancy tests. . . . I used the 2nd test 5 minutes later. Again 'Is this a line? *Does this look like a pink line?*' I'm never buying those cheap, a second pink line means you're pregnant tests again. We all determined it *was* a second pink line and not my imagination" (Baby Corner).

Regardless of how many home pregnancy tests are performed, or what the results are, such tests are not considered authoritative by the test manufacturers, by women, or by their health care providers. Manufacturers advise, women seek, and doctors insist on confirmation by another test done at the doctor's office.[16] As Jordan observed, people generally don't simply "accept authoritative knowledge, . . . but . . . actively and unselfconsciously engage" in "its routine production and reproduction" (Jordan 1993/1978, 152–54).

Even after her significant other confirmed Chrissy's own reading of the test result, she "had blood tests done to verify everything." Jennifer Fisher (2002) tells how, after finding "out I was pregnant" by using a test at home, "I went to the doctor for confirmation. Sure enough, I was pregnant" and Teresa-Lynn, also quoted above, goes on to say "I still wondered if that 2nd pink line wasn't some sort of trick of the light, but a blood test the next day revealed it was a pink line!" (Baby Corner).

Thus, what had in the past been "learning" or "discovering" or "figuring out" that one was pregnant has become a multistep, technologically dependent, diagnostic process. It is not just that missed menses and other bodily changes are no longer considered a reliable source of knowledge.[17] Now, not one, but two and often more scientific tests are undertaken. Home diagnostic kits do not replace doctors' tests; they are just an additional, prior step and represent yet another instance of increasing pregnancy-related consumption (Taylor 2000a, 2000b; Taylor, Layne, and Wozniak 2004). Clearly these tests are profitable for the global[18] pharmaceutical companies that produce them. My local pharmacy offers six brands ranging in price from the pharmacy's own brand at $8.49 for a single test, to $17.39 for a digital name brand, and a whopping $21.99 for a "value pack" of two digitals.[19] All brands offer price incentives to buy multiple tests and some combine regular tests with digital ones in order to entice women to sample this new, more expensive product.[20] One woman recalls how, when she was twenty-eight, she thought "it was a little expensive considering they would also repeat a blood test at the doctor's office. They did not just take your word for it that you were pregnant just from taking the at-home test." As Jordan observed, "The power of authoritative knowledge is not that it is correct but that it counts" (1993/1978, 152–54).

The knowledge such tests provides is of a distinct type; that is, it is biomedical knowledge. The kits are, after all, "diagnostic" tests (Baker et al. 1976; Johnson 1976; Stim 1976). As Nelkin and Tancredi remark in their book, *Dangerous Diagnostics: The Social Power of Biological Information*, diagnostic tests "are widely accepted as neutral, necessary, and benign," when, in fact, "information from tests is not always beneficial or even benign" (1994, 7, 10).

The information home pregnancy tests provide is at once reductionist and universalist. Becoming pregnant (the implantation of a fertilized egg in a woman's body) begins a series of complex physiological changes. These changes are multiple and incremental. Home pregnancy tests fragment, isolate, identify, and measure a single element of these changes; in fact, they measure only a part of one of these elements (the beta-subunit of one hormone). They also universalize—the positive results, that one is "pregnant," suggest that pregnancy is a single thing. But pregnancies are not equal, not even physiologically. A pregnancy test only diagnoses a chemical pregnancy, not a physiological one. Many pregnancies, including those involving a blighted ovum or a molar pregnancy, do not involve the development of an embryo/fetus, but produce hCG nonetheless. One woman explains her "mixed feelings about the early tests because they allow you to get positive results, only to learn it is really a chemical pregnancy or 'early miscarriage.'"

Provision of Privacy

The home pregnancy test means that women can find out if they are pregnant without their doctor knowing. Although this would have been more significant during the era of illegal abortion, it may still be a factor for some women who are considering or want an abortion, especially in small towns. One woman tells of how right after completing college she "didn't want to go to my own doctor for reasons of privacy" and so traveled to "an abortion clinic."

But, as first-person accounts by test users and popular culture depictions of use make clear, home pregnancy tests do not eliminate privacy issues. In fact, the procurement, use, and disposal of home pregnancy tests opens up the possibility of exposure to a greater number and variety of people than a visit to a doctor's office or clinic.

Many of the contributors to the NIH Web site describe the risk of public exposure while buying the tests. One woman recalls "feeling extremely uncomfortable buying the tests. I live in a small town and know a lot of people. I refused to go to Walmart because I just knew that I would run into someone I knew. I went out of town to get them both times." One woman reports that she took a test at 4:30 in the morning because at the time she was "living in an ancient college dorm with community bathrooms" and she wanted to "ensure that she would be alone." She had purchased the test on the Internet and was pleased that "they came in the mail anonymously and the charge on my credit card was also fairly anonymous." One woman recalls how, at age nineteen, she contemplated stealing one because she was embarrassed "to be buying just that one item and didn't have the money to buy anything else."[21] A sophomore in college "remembers feeling like I was sharing a part of my life with the public (or at least, with the people in the grocery store) that I wouldn't have chosen to otherwise."

Several women report having unwanted comments made about their purchase of the test by the cashier. One woman reports, "The guy that rang it up for us asked us if we hope it's positive which I thought was none of his business." Another tells how the clerk assumed she hoped for positive results and "there was an odd discussion in which she was wishing me luck and encouraging me to come back in with the baby and I was dying to get home and get negative results." Another woman reports being annoyed by the intrusion of a pharmacist in Germany who wanted to know why she had come back to the store to buy another brand.

In seven of the twenty-three television episodes involving home pregnancy tests described by Leavitt (2005), the test is discovered in the trash by parents, partners, or friends. These fictional depictions are apparently influencing user behavior. One woman explains, "I threw away the results in the outside trash, not inside, lest I suffer like all those women on tv whose family members 'find' their pregnancy tests in the trash." She also reports being concerned to protect her privacy during procurement. "I was very, very embarrassed to buy the test, but knew I had to know, and would rather face an anonymous drugstore employee than a doctor who would know my name and I hers or his. I used cash so my name would not show up in connection with the test."

Why, and to whom, is privacy important? Most of those who report being concerned about privacy do not want to be pregnant. The nineteen-year-old who contemplated stealing the test "cried alone for two hours" when the result was positive. The woman who preferred an anonymous drugstore employee over her doctor was single, living at home with her mother, between jobs, "absolutely poor," and without health insurance when she woke up naked in bed with a man one morning after a party. The test was positive and she had an abortion. The woman who went to Walmart was living alone in the Midwest. She "definitely did not want to be pregnant" and was "elated with the results of negative. I remembered almost crying I was so happy and thanking God." And the woman who purchased the tests on the Internet and took precautions to be alone in the dorm bathroom while using it "definitely wanted the result to be negative, since I was still in college and unmarried." Another woman who "desperately wanted [the results] to be negative" went to a pharmacy "where no one knew me!"[22]

In contrast, women who wish to be pregnant often report being eager to announce the news of their pregnancy. One married woman tells of feeling "proud as I stood in the test aisle and took a great deal of time reading all of the boxes to see which one would give me the most accurate result (and she contrasts this with her feelings of embarrassment twelve years earlier when she was single and possibly pregnant by someone other than her boyfriend). Sometimes women employ a used test stick as an announcement. One woman tells of how on Valentine's Day she "secretly took a home pregnancy test and got a big positive. I presented the test to my husband as a V. Day gift." A man tells of coming downstairs one morning where he was greeted by his wife "with a big smile and a home pregnancy tester wrapped in a bow." On a 2004 episode of *The L Word*, a popular lesbian television drama in the United States, "Tina surprises Bette with the news that they're going to have a baby by setting the dinner table for three and putting the positive pregnancy test at the third setting" (Leavitt 2005). Others report keeping the used stick as memorabilia: "[T]he sticks became treasured items and are on the first page of each child's baby book." Another tells that she wishes she "had saved the wand, as my first memento of my son, announcing his future arrival."

Even women who wish to be pregnant sometimes report resenting their loss of privacy while purchasing, however. "I remember not wanting anyone to see me purchasing it, but only because until I knew I was pregnant I did not want anyone else to know that we were trying." Another recalls being "a bit nervous purchasing the test, only because I had not yet told any friends that I was trying to get pregnant so I didn't want to have to answer any questions if they saw me with the test," and a professor who was trying to get pregnant reports worrying about running into her students at the store.

Faster, Faster, Faster

Advertising for home pregnancy tests stresses how fast[23] their product can detect a pregnancy. In 2005, when I started research on this subject, the CVS brand "Early Pregnancy Test" was, in fact, one of the later tests in that it promised results from the first day of a missed period. "Clear Blue," made by Unipath in the United Kingdom, and "First Response," made by Armkel in California, both boasted "earliest results," "results 4 days sooner than other leading brands." Armkel also marketed Answer, their less-expensive product, saying it could provide results three days before an anticipated period.

By 2008, all six brands for sale in my local pharmacy boasted test results "5 days sooner"—that is, than a missed period—and "over 99 percent accurate" in large print on their packages even though the accuracy rates for the "5 days sooner" diagnosis is only 53 percent.[24]

Does such early diagnosis really benefit women? Certainly not if there is only a fifty-fifty chance of getting an accurate result at the earliest date advertised by the test makers. Rather than waste their money, women might just as well flip a coin.

How about the tests when used after the day of a missed period, at which point they are 90–99 percent accurate?[25] Early supporters of the test argued that "earlier diagnosis of pregnancy" would result in "earlier prenatal care and earlier abortion, thereby contributing to better maternal health" (Baker et al. 1976, 167). But how early is optimal, and how early is early enough? And do the tests actually deliver on this promise?

Earlier Prenatal Care?

Earlier prenatal care has been touted as a benefit of home pregnancy tests both in terms of maternal and fetal health. In 2005, the e.p.t. Web site instructed users

Selection of home pregnancy test kits in a pharmacy in Troy, New York, 2008. Photo by Linda Layne.

who test positive that "[e]arly prenatal care is important to ensure the health of you and your baby."

Improving maternal and fetal health is a worthy goal. The United States continues to rank near the bottom of developed nations in terms of maternal/fetal health (U.S. Department of Health and Human Services 1999, 20). In 1995 the risk of a Finnish child dying in infancy was 48 percent lower than that in the United States (U.S. Department of Health and Human Services 1999, 20). Class and racial differences loom large in this regard. In the United States, infant mortality is more than twice as common for black infants than for white (U.S. Department of Health and Human Services 1999, 21). Maternal mortality is three to four times higher for black women than for whites. This difference is most apparent in pregnancies that do not end in a live birth because without proper medical care, women can die from ectopic pregnancy, spontaneous and induced abortions, and gestational trophoblastic disease (Cunningham et al. 2001, 8–9). Much of this disparity is thought to be the result of "poor availability of medical care for minority women," which in turn is linked to "the erosion of health-care safety nets for the uninsured" (Cunningham et al. 2001, 10–11).

Better outcomes are found in pregnancies where there is "early and consistent prenatal care" (Winston and Oths 2000), probably because this enables detecting and managing serious problems such as gestation hypertension and diabetes, and providing patient education about risky health behaviors such as smoking[26] and drinking. Thus, providing earlier prenatal care to the underserved should help reduce the disproportionate rates of pregnancy loss and infant mortality and morbidity, and maternal mortality and morbidity among poor women of color. But home pregnancy tests are not the way to achieve this. Home pregnancy tests have been available in the United States now for more than thirty years, yet in 2005 nearly one-fifth of pregnant women in the United States still did not receive prenatal care in the first trimester[27] (Martin et al. 2007).

For middle-class women, it is unclear whether home pregnancy tests have led to earlier prenatal care and, if so, whether this has been beneficial. Because of the high rate of pregnancy loss in the earliest weeks, some obstetrical practices will not begin prenatal care until after a heartbeat has been seen; that is, not until six to eight weeks. A preconception visit at which potential problems like hypertension, diabetes, and Rh factor can be addressed; prenatal vitamins be prescribed; and information on spontaneous abortion be provided would, in my view, be more beneficial than prenatal care during the earliest weeks of pregnancy (Layne 2006a, 2006b, 2007).[28] For women who are already "taking care of themselves," there may be physical and psychological advantages to taking a more "wait and see" attitude about early pregnancy.

One woman who reports having used "*many* home pregnancy tests" says she "has mixed feelings about the early tests because they allow you to get positive results, only to learn it really is a chemical pregnancy or 'early miscarriage.'" Another woman who submitted her story under the name "needababybadly" tells how she "bought 5–6 different kinds just to make sure" and then "purchased loads more because I had waited so long to be able to test." Her results were "*positive, positive, positive*" because she was carrying twins but she ended up losing this pregnancy.

I have described elsewhere the costs of earlier acknowledgment of, and investment in, a pregnancy for the nearly one million U.S. women whose pregnancies end in loss each year (Layne 2003). Use of the home pregnancy test means that women who in the past would have been spared the experience now must deal with a loss, and do so in a culture that denies and belittles this experience. Because miscarriage is so common, and is becoming increasingly common as a result of increasing maternal age, environmental toxins, and in vitro fertilization, this subgroup of test users is important to acknowledge.

Earlier Abortions?

In the United States, abortion is legal until twenty-four weeks' gestation. Thus, from a legal point of view, any time before the twenty-fourth week is early enough. But there are a variety of procedures, and these vary depending on how far along one is and where and by whom one is being seen. Planned Parenthood offers two choices up to nine weeks—medicated and vacuum aspiration—and vacuum aspiration after nine weeks. Although practice varies from clinic to clinic, overall, one has more options during the first trimester (first twelve weeks) than after, and there may be some health advantages to first-trimester abortions, but these are not entirely evident and there is no magic cutoff point after which things change. Thus, one answer to the question of how early is early enough is early enough for a first-trimester abortion.

But let us work from the other direction too. Fifteen to 20 percent of confirmed pregnancies end in spontaneous abortion (the rate is even higher for young teens, a population for whom pregnancies are often unwanted), and because most of these occur in the first weeks of pregnancy, might there not be advantages to testing somewhat later, for example, two or three weeks after an expected period?[29] Furthermore, because the very highest rate of miscarriage occurs between fertilization and the first missed menses (accounting for an additional 10 percent), does it not make sense to wait at least until the first missed menses?

Baker et al. (1976, 167) recognized that false positives "could cause the woman unnecessary psychological stress and expose her to the expense and potential risk of an unnecessary . . . abortion." Earlier and earlier diagnosis subjects women to these same risks.[30]

HOME PREGNANCY TESTS AND PREGNANCY LOSS

Home pregnancy tests have also changed the experience of pregnancy loss. Earlier and more intensive medical management of pregnancy encourages earlier and more intensive social construction of fetal personhood in wished-for pregnancies and to the view of pregnancy as something that can and should be controlled. In the past, as Duden tells us, physiological changes in a woman's body were "signs and intimations of" a pregnancy; one could never be sure that one was going to have a baby, "it remain[ed] a hope" (1993, 9). But during the last quarter of the twentieth century, "hope . . . dissolved into expectations that can be managed at will" (Duden 1993, 10). The innocent-looking home pregnancy test is in fact one of the technologies that has contributed to this epistemic shift. Women whose pregnancies end in loss suffer as a result (both at the moment of loss and during subsequent pregnancies).

Home pregnancy tests feature prominently in narratives of loss and in accounts of pregnancies that occur after a loss. Precisely because the tests encourage prenatal bonding in women who wish to be pregnant, they often provoke strong negative feelings or deep ambivalence during subsequent pregnancies. Heather Gail Evans-Smith (2002) recalls, "The test is positive and yet I am not overjoyed, But in pain, in fear of all the what ifs/The memories of before burrow through my brain/I sink to the bathroom floor/EPT test still in my hand/It is positive I whisper to the walls/It is true I am with child . . . I shake my head in despair/ . . . I am pregnant and don't know what to say/I don't know what to think/I don't know what to feel/ . . . As I lay on the cold tile/ Grasp the test in my hand /I wonder if this time . . . I will hold a baby in my arms" (Evans-Smith 2002). Jennifer Fisher (2002) explained her mixed feelings upon discovering with a home pregnancy test that she was pregnant again, following a miscarriage at six weeks' gestation the previous year, "I was excited but also scared. I was scared that I was going to lose this one too."

During a subsequent pregnancy, home pregnancy tests are also sometimes used as "appropriated technology"; that is, for a purpose not anticipated by the manufacturers and marketers (Eglash et al. 2004). Some women seek more frequent medical testing during a subsequent pregnancy, not to reassure themselves, but to prepare for the next anticipated loss, and some of these women have figured out that pregnancy testing can be used to this end.[31] One woman explains how during a subsequent pregnancy she "wanted an hCG then. I wanted an hCG in a week. You know, it was like I was going to track this one and find out if there is any demise coming" (Cote-Arsenault and Marshall 2000, 482). A woman who was scared she would miscarry again tells of how she had several positives over a few days but because of her concerns a friend suggested "I get the digital because seeing the word might make it feel 'real.' Well sadly it said, 'Not Pregnant,' and I started to miscarry the next day."

WHO DO HOME PREGNANCY TESTS BENEFIT?

Home pregnancy tests clearly benefit the pharmaceutical companies that manufacture them. In 1997 home pregnancy tests in the United States alone accounted for $206 million dollars in sales. Clear Blue Easy (Unipath Diagnostics, a U.S. subsidiary of Unilever), which at that time held a 6 percent share of the market, hired David Lynch to do a seven-million-dollar ad campaign consisting of one thirty-second commercial and two fifteen-second ones (Charry 1997). The ads were based on "an understanding of female psychology . . . we looked at the emotional process involved."

Despite the appearance of giving women more autonomy, home pregnancy tests, in fact, create a new technology/consumer/pharmacological dependency. These tests are also being marshaled to promote other such dependencies under the guise of facilitating women's control over their own bodies. Ovulation predictor kits are manufactured by the same companies that make home pregnancy tests. The insert in the First Response pregnancy test advises women who want to be pregnant but got a negative result to use one of their "First Response Easy-Read Ovulation Test kits, an at-home, one-step" hormonal test so "you can . . . get pregnant sooner." It also includes coupons for buying their ovulation kits and more pregnancy tests. The Evatest brand pregnancy test, purchased for me in Uruguay in 1980, also advertises its "Evaplan-Ovulation test in one step" that "predicts in just three minutes the most fertile period."

In addition, home pregnancy tests are being used to promote menstrual-suppressing birth control pills (Aegnst and Layne, this volume). Proponents of menstrual-suppressing birth control pills argue that the week of placebo pills in the classic twenty-one/seven schedule of oral contraceptives was included to reassure women that they were not pregnant, but now, because that is "easily handled . . . with home pregnancy urine dipsticks," this obstacle to the spread of extended-regimen pills has been eliminated (Thomas and Ellertson 2000, 923). Thus, home pregnancy tests are understood to pave the way for "continuous pharmaceutical management" of menstrual cycles (Thomas and Ellertson 2000, 923).

Home pregnancy tests also benefit the stores that sell them. A 1991 article in *Drug Store News* reported that pregnancy test sales in drug stores had "surged 450 percent from 1982–1989. Ovulation predictor kits have continued to grow an average of 70 percent since 1986." They instruct drugstores how to optimize "your share of this *high profit margin* category" by increasing shelf space and locating the tests in "one convenient Family Planning Center" that will increase "the likelihood of multiple purchases among companion products" (Drug Store News 1991, emphasis added).

Do they benefit women too? We do not have the data needed to answer this question, and the fact that we don't opens up another important series of questions. Home pregnancy tests benefit women if they lead to better pregnancy outcomes or safer abortions.[32] Whether or not they do are unanswered empirical questions. In order to know what the actual health benefits or deficits of home testing are, one would also need to know whether women who do not wish to be pregnant actually get earlier abortions. If so, how much earlier, and are there significant health benefits because of the time difference? Do women who wish to be pregnant actually get earlier prenatal care? If so, how much earlier, and does the time difference actually account for improved health of women and/or their embryo/fetus?

Earlier abortions and earlier prenatal care pertain to women who are pregnant. What are the benefits and harms of home testing for those who are not? A study of teen users suggests that the availability of home pregnancy tests is reducing the ability of health care providers to offer family planning counseling to at-risk teenage girls (Shew et al. 2000).[33] This study found that teenage girls who used home tests were less likely than the control group to be using effective birth control methods, and because most of those who got negative results did not follow up with a clinician, the kits reduced "provider access to at-risk youth for pregnancy prevention counseling" (Shew et al. 2000).

One of the advantages the home pregnancy test was expected to provide was "allowing a woman to be the first person to know that she is pregnant" (Baker et al. 1976). This view ignores the fact that in the past women were the first people to know, though via different means. By externalizing an internal bodily state, home pregnancy tests make this information readable not only to the woman but also to others. Home pregnancy kits greatly extend the number and types of people who might know that they might be pregnant, including members of the public (cashiers and people who witness them buying the product or find the test in their home or in the trash). It also makes this information more readily available to the men in their lives. Although most women who shared their stories on the NIH Web site welcomed their sexual partner's involvement, one report offers a sobering reminder that such involvement is not always a good thing. A twenty-four-year-old recalls using a test when she was seventeen, still living at home with her family, and involved in an abusive relationship. Her boyfriend "sat down in the middle of" a store "and opened up the package. He laid out the instructions all over the floor and told me to go pee on it and bring it straight back to him. He told me he wanted to make sure I didn't tamper with it. . . . I did what he asked . . . I wasn't allowed to see the test, and he kept it with him." Even in healthier relationships, involving male partners can prompt controlling behavior. In Swain's novel, Eddie hands Lemon a folic acid pill with a glass a milk before she even takes the pregnancy test, explaining "calcium . . . it's good for you" (2004, 62). And when she asks for tea while she is taking the test, he says "Okay, herbal only. No caffeine" (Swain 2004, 69).

A FEMINIST HOME PREGNANCY TEST

Being able to tell whether one is pregnant or not is clearly of value to women. There are several ways to gain this knowledge, however, including several that do not involve the purchase of a one-use, disposable product.

Before the advent of early pregnancy testing, women learned they were pregnant by noticing changes in their body. According to a 1973 study of the thirty-three women who came to the Feminist Women's Health Center in California because of a possible pregnancy, twenty-eight were "convinced" that they knew whether they were pregnant based on changes in their body (breast changes being the most frequently mentioned, followed by missed menses, nausea, fatigue, and many idiosyncratic symptoms that women reported having experienced during previous pregnancies). Of these, all but one were correct (the only one who mistakenly thought she was pregnant had been taking hCG as an obesity treatment, which causes women to feel pregnant and may result in a positive pregnancy test) (Jordan 1977, 21). In addition, four of the five who "suspected" they were pregnant but weren't entirely sure were in fact pregnant (again the only one who wasn't had been taking hCG for weight loss). In other words, competence to accurately self-diagnose early pregnancy based on embodied experience was "massively present" (Jordan 1977).[34]

Many of the women who contributed to the NIH Web site also reported that they possessed self-knowledge of what the test would show. For instance, one woman said, "I kind of had a feeling of what the results would be. My reaction was just confirmation. Kind of like I knew it." And another woman reports, "I was so hungry all of a sudden, I knew something was up." Another woman shares that her mother told her how back when the tests took so long to use, "she just sat there . . . for hours in suspense waiting for the result even though she already knew in her heart." One woman who did not want to be pregnant but was when she was seventeen remembers "the feeling of dread waiting for the color to turn. . . . I knew I was pregnant but that damn little test would confirm all my worst fears." Another woman "fresh out of college" writes "I knew I was pregnant before I even took the test." This woman also says she used the tests other times when she knew that she wasn't pregnant but her period was late. The nineteen-year-old who contemplated stealing the test explains, "I knew it would be positive. My gut already told me. I still cried forever."[35]

Of course, not everyone recognizes when they are pregnant, even if they have been pregnant before.[36] One woman who turned out to be pregnant was thirty-two and at home with eighteen-month-old twins that she had used assisted fertility technologies to conceive. She explained that her husband bought the test for her because "I didn't think I could be pregnant." A forty-five-year-old woman who also was surprised to learn she was pregnant said, "I thought I knew my body well; you think you've got it covered. I was tired a lot and my period was light, but I didn't even consider it . . . until a colleague asked, 'Could you be pregnant?'" (Boston Women's Health Book Collective 2005, 383).

As *Our Body, Ourselves* explains, "[C]ulture and family upbringing may influence how we interpret the changes in our body. Many of us think we have digestive problems, stress, or the flu. Some of us have taken so many risks without conceiving, or tried for so long without luck, that we think we are infertile. For some of us, it is unthinkable, and we just do not accept the signs. For others, we do not want to make decisions about the pregnancy so we wait until we are so far along that our options are limited" (Boston Women's Health Book Collective 2005, 384).

Nevertheless, this valuable self-knowledge should be recognized and cultivated as an important feminist means of detecting a pregnancy, one that offers several advantages over home testing. Unlike the false claims of home pregnancy tests, this method does enhance women's autonomy. Similarly, it actually does maximize the potential for privacy—no one needs to know until the woman deems it appropriate. Furthermore, it doesn't cost anything. If the pervasive adoption of home pregnancy testing has resulted in women losing the skill to detect their own pregnancies, this is something that can and should be regained.

Another feminist method to determine if one is pregnant is by doing a pelvic self-exam or helping a friend do one. One contributor to the NIH Web site recalls having had a "plastic speculum around from the do-it-yourself pelvic exam days" that she used to diagnose her "first pregnancy by observing the changes in my cervix." The pelvic exam was mentioned as one of four methods in *Our Bodies,*

Ourselves in the edition that came out after the advent of home pregnancy tests but is not mentioned in the most recent edition. This appears to be another valuable skill American women have lost since home pregnancy tests became available, and one that can be regained.

If we adopt a liberal feminist stance and concede that more choices are better and that commercially available, hormone-based kits will hold some benefit for some women under some circumstances, then we should work to improve them to better serve women. At present, even though home pregnancy tests measure hCG levels, they do not reveal this level to the user. Even the expensive digital ones do not actually tell women what their hCG level is, only whether it is high enough to indicate a pregnancy is likely.[37] The actual hCG level (especially if tracked over time by using repeat tests) can be an important indicator of many things, including whether a pregnancy is likely to end in miscarriage, whether it is likely to be a multiple gestation (twins, triplets, etc.), or whether it is likely to be an ectopic pregnancy. Although there has been a significant drop in the death rate for ectopic pregnancy since the 1970s, about forty to fifty women die each year from ectopic pregnancy in the United States. In early pregnancy, hCG levels generally double every seventy-two hours, and plateauing levels or an abnormal rise often indicates ectopic pregnancy. Putting this information in women's hands might save lives (http://www.advancedfertility.com/ectopic.htm).

In addition to this "hardware" change, there are improvements in the "software" that would help make these tests better feminist information technologies by providing information women need to make informed choices. The six brands I examined in 2008[38] included inserts that, along with directions and disclaimers, answered a pair of "what should I do if" questions. If the "test says I'm not pregnant," the inserts advise using another of their tests in a few days and "if your period hasn't started in a few days, go see your health care professional." If the "test says I'm pregnant," the inserts chide, "Remember, this is not intended to replace your doctor's diagnosis. . . . See your doctor to confirm you are pregnant" and "who can advise you on what steps you should take next." Only First Response and its less-expensive sister brand, Answer, gave information on how to "increase your chances for a healthy pregnancy."[39]

None of the brands mention abortion, or even acknowledge that some of the users who are pregnant will want to keep the pregnancy and others will not.[40] Nor do they mention the possibility of pregnancy loss.[41]

Rather than simply referring women to a doctor, a feminist pregnancy test would provide women with information on miscarriage, elective abortion, best practices during the early weeks/months of a desired pregnancy, and resources for getting more information. It would also provide information for women who discover they are not pregnant and are relieved about how to avoid such a scare in the future, and for those who are disappointed, about what this might mean and, if appropriate, direct them to infertility resources. A good starting place for modeling feminist "software" (Hardon, this volume) to go with the existing hardware" of home tests can be found in the 2005 edition of *Our Bodies, Ourselves* (pp. 384–88). The Boston Women's Health Book Collective discusses what to do if you were afraid you were pregnant and turned out not to be (find out if there are contraceptive methods better suited to you; how to get help if you are being sexually abused; how/where to get emergency contraception; talk with a family planning clinic). For those who test positive and need to figure out what do, they discuss insurance coverage, who to turn to, family pressure and strong-willed partners, poverty, the special challenges of teen pregnancy, abortion, carrying to term, parenting, foster care, adoption, and where to get more information. This section of the book is about as long as the current product inserts; hence, such information could easily be provided. Perhaps the Boston Women's Health Book Collective could partner with a test maker (as one brand has done with the March of Dimes).

In any event, I would advise using home pregnancy tests sparingly, and when considering using them to be mindful of the ramifications. A better solution, in terms of ensuring safe abortions and prenatal care beginning during the first twelve weeks of pregnancy, would be to do as the California State Department of Health did in 1976 in response to the advent of home testing (Cunningham 1976)—that is, to support legislation that mandates free testing to all women at publicly supported family planning clinics that would also provide confidential qualified counseling and publicly funded resources for abortion or prenatal care.[42] For women who feel the need for early detection, I advise using existing resources like Planned Parenthood where the test is free, confidential, and proximate to other pregnancy-related resources.

CONCLUSIONS

As we have seen, the home pregnancy test has been massively adopted during the past thirty years as "the right tool for the job" (Casper and Clarke 1998) of figuring out whether or not one is pregnant. Yet despite the rigorous feminist critique of the medicalization of pregnancy and birth and of many of the new reproductive technologies, there has been almost no assessment of the home pregnancy test. Interdisciplinary feminist scholarship on new reproductive technologies has focused more on high-tech, costly technologies, including in vitro fertilization, prenatal testing, fetal imaging, and fetal surgery. In STS too, low-tech products or solutions like the home pregnancy test tend to be championed but rarely studied (except in the South under the rubric of "appropriate technology").[43]

The home pregnancy test represents a classic case of "technological somnambulism" (Winner 1986). We were sleepwalking and thoughtlessly accepted a new technology that "reconstituted the conditions of [our] existence" (Winner 1986). This seemingly simple little technology has changed the way women experience infertility, pregnancy, abortion, and pregnancy loss and these changes have not always been in ways that benefit women compared either to the preexisting status quo or with potential alternatives.

Probably one of the reasons that it was accepted with so little thought is that it seems so innocuous and unimportant. In the United States at least, another reason has to do with the particular moment in history at which these tests entered the U.S. market—right on the heels of decades during which abortion was illegal and unsafe.

The case of pregnancy loss highlights the need for systematic feminist technology assessment. It also points to the need to refine the criteria by which feminist technologies are judged. If we use the criteria that have been established among the international women's health activists for evaluating proposed forms of birth control—that is, safe, effective, is administered by the woman, and has no irreversible consequences (Hardon, this volume)—the home pregnancy test looks like a winner. But as we have seen, a fuller understanding of the social circumstances under which women buy, use, and dispose of these technologies and the consequences test use has on women's sense of self, relationships, options, and actions are essential. Case studies like these provide a fruitful opportunity to develop a more close-grained model for designing, promoting, and evaluating feminist technologies.

NOTES

Thanks to G. J. Barker-Benfield, Maureen Gallagher, Shirley Gorenstein, Deborah Johnson, Lynn Morgan, Liz Roberts, and the anonymous reviewers for many important suggestions. I am also grateful to Council on Anthropology and Reproduction members Nicole Berry and Lynn Morgan and Lynn's husband Jim Trostle who kindly took the initiative to acquire tests for me while traveling, and STS graduate students Richard Arias and "Denver" Xiaofeng Tang for translating them.

1. They are not so easily available in all other countries. A forty-two-year-old American woman tells of using a home pregnancy test in 2003 in the Philippines where she and her husband had recently moved. She couldn't find one in the pharmacy, so she asked but felt embarrassed doing so because it is "a very Catholic country" and she "made a point of

making sure [the clerk] saw my wedding band!" She learned that there is only one brand sold in the country and that they keep the tests "under lock and key, although they are relatively inexpensive at . . . about $1.90."

2. These accounts can be found at http://echo.gmu.edu/old/nih/responses.php. The pregnancy Web site is www.thebabycorner.com. The pregnancy loss support newsletters are the quarterly newsletters (1981–2004) of UNITE, a regional group with about ten support groups serving Pennsylvania and New Jersey and the six annual newsletters (1984–2004) of SHARE, the largest pregnancy loss support organization with groups throughout the United States.

3. One contributor to the NIH Web site tells how she performed the test for her aunt in 1959 when she was working at a pathology lab as a medical technologist. She recalls, "[T]he frogs were kept in the refrigerator until use so they would be subdued enough to handle. A small sample of the patient's urine would be injected under the skin of the male frog and after several hours at room temperature, the frog was held in a way to press his bottom against a glass slide which was then examined under a microscope for sperm. . . . I remember the thrill of seeing the sperm under the microscope and her [aunt's] scream over the phone when I told her she was pregnant."

4. The test was first used by clinicians treating hCG-secreting tumors because the radioimmunoassay showed whether chemotherapy treatments had worked.

5. In 1972 the FDA had one manufacturer of a home pregnancy test that measured estrogen levels in women's urine remove it from the market because it was so inaccurate (Baker et al. 1976); in 1975 a federal judge ruled that the FDA did not have the power to restrict the sale of the kit nor to require the company to demonstrate that it was safe and effective because it was not a "drug" (Johnson 1976). Nor did it qualify as a medical device because pregnancy is not a disease (Johnson 1976). The new digital models, however, must have Federal Communications Commission approval because they can interfere with reception.

6. These were e.p.t. (initially standing for "Early Pregnancy Test" and later known as the "Error Proof Test"), Predictor, ACU-TEST, and Answer.

7. A woman who took the test in 1995 recalls that "the test came with two test tubes (one full of liquid), a little cup, and an eye dropper. Simpler tests were available, but being an impoverished grad student I bought the cheapest one. It took about ½ hour total to read and follow the instructions."

8. Several young women remember hearing from their mothers that "the tests back then were horrid." One woman who was born in 1983 fondly recalls how her mother would tell her about using a test in 1982 and "within ten minutes there was a dark ring and she knew she was having me!"

9. It is worth noting that no users are pictured on the packaging of the products reviewed in the United States, China, Uruguay, or Argentina. The only woman pictured on any of these is found on the Evatest from Uruguay and she represents not a user, but an employee who is available to answer questions by phone. One woman who contributed to the NIH Web site reports that in the Philippines, "the packaging was too full of drawings of cute babies, ribbons and flowers to convey much information"; the package of a test

she bought in Japan had a drawing of a happy woman on the box who she interpreted could be "smiling because [either] she is or is not pregnant," whereas the "Filipino brand's package made it seem very clear that there was only one result that a woman would want to see when taking the test."

10. In addition, some users are not sure whether or not they want to be pregnant.

11. The inserts of the six tests I examined were in English and Spanish. Answer brand, which is the bargain brand made by the same manufacturers as First Response, also has both languages on the box; none of the others did.

12. The David Pregnancy Test, purchased in Beijing, China, in 2008, instructs users who have positive results to "please see the doctor to confirm the pregnancy, the doctor will give you further advice to guarantee the health of you and your fetus." There is no mention of the fact that pregnancy loss may occur regardless of what advice is followed.

13. Mothers and stepmothers also sometimes report buying them for their teenage daughters.

14. In 1976 a letter was published in the *American Journal of Public Health* by a member of the British Institute of Medical Laboratory Sciences regarding his concerns about the "use of such kits by non-technical staff such as nurses, receptionists, and clerical employees in health centers and in private practice, and especially . . . by the patient herself. [Because] such users do not in general have sufficient training to detect malfunctions in these materials and not understanding the complex nature of the reactions involved, [they] often try to modify the procedure to suit them, with predictably catastrophic results" (Entwistle 1976).

15. Leavitt describes many portrayals of test use on television, of which this is "the most culturally significant" one. She notes that the test's "very presence on so many different kinds of television shows and movies geared at different audiences hints at its cultural relevance. The pregnancy test is an example of a home healthcare product that is familiar to millions of people."

16. In the past, the tests done in doctor's offices were typically blood tests—that is, different than the home tests—but one woman recalls after having gone to Planned Parenthood, "I was kind of upset when I realized they really use a test similar to the over-the-counter tests I had already bought and taken."

17. It is worth noting similar changes in diagnosis by physicians. As Nelkin and Tancredi note, "for most of medical history, diagnosis has been limited to measuring the external manifestations of disease. . . . The general physical examination, for years the standard forum [for diagnosis], relied on the sense of the physician. . . . Few tools were available to confirm the physician's intuitive impressions about a patient's condition" (1994, 20).

18. The CVS brand tests are "made in the U.K., assembled in China, packaged in the U.S.A."

19. Several contributors to the NIH Web site comment on the issue of price. A twenty-two-year-old explains that she bought the least expensive one because she was broke. Another woman bought four tests at a "99¢ store." Another woman recalls that at age eighteen she "went to the store after work, walked up to the section, sat in dismay at the prices for a bit," and ended up buying the store brand. A nineteen-year-old bought two name-brand

tests but was "upset by the cost." One woman's husband "commented on the high cost of the test." Another explains she "bought a generic brand because I didn't want to spend the extra money if I really wasn't pregnant," and muses, "Crazy, huh?"—the implication being that for a child, no expense would be spared, but for a false alarm, there was not such sense of investment. Apparently, by 2003 the pregnancy test kit sales were flat, with $138.7 million in drug store sales; in 2002 women bought twice as many of the cheaper (55 percent less) store-brand tests than the leading brand, e.p.t. (Johnsen 2003).

20. Inverness received approval to market a digital pregnancy test in May 2003 and by October the company had sold about $1 million worth in drug stores in the United States (Johnsen 2003). Since that time, all the major brands have followed suit with digitals of their own.

21. Another reports how in her twenties she was tempted to "try and 'hide' it with lots of other purchases in the basket as one might in the case of condoms or even sanitary products or anything related to reproduction." (See Vostral for a discussion of the stigma attached to being a menstruating woman). One woman reports that she found buying the test easy, but explains, "I'm really comfortable purchasing 'personal' items like tampons or condoms."

22. A nineteen-year-old reports that she drove "3 cities over" to buy the test and was "still embarrassed and looked down through the entire transaction." She wanted a child but was relieved by the negative result because she was "still so young" and "my life didn't have to change."

23. Speed of use is also a featured element in the marketing of these kits. Packages stress how quick they are to use—"hold stick just five seconds in urine," "results in 1 minute," "one quick step."

24. Two of the six brands printed a table with these accuracy figures on the box and one of these (e.p.t.) actually advises, in bold print, that women "wait until the day of your expected period." These results (53 percent accuracy four days before; 74 percent three days, 84 percent two days, and 87 percent one day before the first day of a missed period) have not improved since at least 2005.

25. Makers claim a 99 percent accuracy if used at this date but a 2001 article in the *Journal of the American Medical Association* addressing the "natural limits of pregnancy testing" accuracy disputes this claim. A pregnancy cannot be detected before the blastocyst implants and the timing of ovulation does not always occur in the middle of a menstrual cycle but may occur near the end. The authors argue that this means that "the highest possible accuracy for hCG-based tests is 90 percent on the first day of the missed period (Wilcox et al. 2001). A 1986 article that compared makers' claims with actual in-home use of 109 women found even greater discrepancy (Doshi 1986).

26. The rate for smoking during pregnancy is still about 10 percent (and is especially prevalent among women eighteen to twenty-four years of age) even though it is associated with low birth weight.

27. The rates of prenatal care use among "groups which historically have had lower levels of care: non-Hispanic black, Hispanic, and Asian women" have increased since the early 1990s and this is attributed to the expansion of Medicaid for pregnant women (Martin et

al. 2007, 17). However, these groups still remain "more than twice as likely as non-Hispanic white women to receive" no care or only care beginning in the final trimester (Martin et al. 2007, 17).

28. Landsman (2009, 24–25) cautions, however, that the new trend toward preconception care increases the scope of "mother blame" for any less-than-perfect reproductive outcomes.

29. In the early twentieth century, late menses were not considered "reliable indicators of pregnancy" and "cycle restorers" were sold through the mail and sometimes even off the shelf . . ." (Maines 2003, 111). I am suggesting that even in an era of legal abortion, there might be some advantages to women of this level of ambiguity.

30. Furthermore, the medicalization of pregnancy and growing ideology of fetal personhood that home pregnancy tests support has changed the way abortions are experienced by women. A positive result on a home pregnancy test leads one to "conjure up a fetus, and with it the abstraction 'life'" (Duden 1993, 53). Duden (1993, 54) describes a German family planning center that performs first-trimester abortions only. The staff had always let "the patient glance at the tray containing the product of the suction process. During the last two or three years, the reaction of women . . . changed. Many now see in this bloody mass the face of a child."

31. Another example of an unanticipated use is that some use the tests to bring on their menses. For example, a woman tells of taking the test with a friend when their periods were late: "[W]e were very relieved to see we were not pregnant. I think our bodies were too because we both got our periods within a day or two of taking the test." Another woman tells of using the tests when her period is irregular. "I know that I am not pregnant but the negative result relieves my stressed system and my cycle usually starts back up." Another says of her test use in her twenties, "[U]sually I knew deep in my heart that I wasn't pregnant, but I wanted to be sure. It would help get my period started! . . . I was always relieved when it was [negative] and then my period would start." Apparently none of these women considered that if they had only waited another day or two, their periods would have come (without the needless use of a test) and their worries would have been allayed.

32. When the tests were first introduced, some believed a benefit would be that it would encourage women to go see their doctors because, according to them, "Some women prefer to wait until they are sure they are pregnant before visiting their physician" (Baker et al. 1976). Others feared that the kits would have the opposite effect. The chief of the Maternal and Child Health Department of California feared they might discourage women from seeking medical care (Cunningham 1976). As far as I can tell, no studies have been done to ascertain what the effects actually are.

33. Similar concerns are being voiced in the current debate about whether to legalize over-the-counter HIV tests for home use. Although a recognized advantage is that the privacy afforded at home would encourage more people to get tested, the concern is that if people test at home, they will be less likely to get counseling and to understand how to avoid risky behaviors that might infect others (Leland 2005).

34. Other examples can be found in *Our Bodies, Ourselves* (Boston Women's Health Book Collective 1976). See also Davis-Floyd and Davis (1996) on the status of intuition.

35. In the opening scene of the movie *Juno,* a female customer tells Juno how easy it is to tell if one is pregnant based on bodily changes. She uses improper English and appears to be shoplifting while Juno is talking to the clerk. Thus, one might infer that only ignorant, uneducated lowlifes who cannot afford pregnancy tests would use this outdated form of diagnosis.

36. Some report that they did not suspect but others "knew." Some tell of their husbands or coworkers knowing before they did and buying the test for them based on that "knowledge."

37. The JSJ test purchased in Beijing in 2008 offers four results instead of the standard two found on tests in the United States: "Negative: if only the contrasting line appears, there's no pregnancy. Weak positive: if two lines appear, and the color of the testing line is much lighter than the contrasting line, there's probably a pregnancy; please use the morning urine to retest after two days. Positive: if two red lines appear, and the testing line is clear, there's a pregnancy. Ineffective: if no line shows, the test is ineffective, please use new test paper to retest."

38. CVS Early Result Pregnancy Test; CVS One Step Pregnancy Test; First Response Early Result Pregnancy Test; Answer Early Result Pregnancy Test; Clearblue Easy Digital Pregnancy Test; e.p.t. Pregnancy Test.

39. They advise: (1) maintain a well-balanced diet; (2) stop smoking; and (3) do not drink alcoholic beverages. They have partnered with the March of Dimes.

40. Both of the Chinese tests I had access to use the term "early pregnancy," the character for which (zao yun) also refers to "premature pregnancy," which usually indicates a girl who gets pregnant before marriage.

41. Professor Nicole Berry from British Colombia reports that neither of the tests she bought there mentioned abortion even though abortion is provided through the national health service and the topic is not as politicized there as it is in the United States. The test made by Mansfield advises those who test positive, "You should see your doctor. The earlier you see your doctor, the better it is for you and your baby." And the First Response test gives some tips for a healthy pregnancy, then concludes that "you need to see your doctor immediately for proper care and nutrition counseling."

42. This is what Cunningham and his colleagues at the California Department of Health did in 1976.

43. A similar observation was made by Casper and Clarke with regard to the pap smear, one of the "supposedly simple technologies that are widely used" but understudied in the field of science and technology studies (1998, 256).

References

Baby Corner. www.thebabycorner.com/journals. (Accessed April 20, 2002.)

Baker, L. D., L.W. Yert, M.C. Chase, E. Dale. 1976. "Evaluation of a 'Do-It-Yourself' Pregnancy Test." *American Journal of Public Health* 66: 166–67.

Boston Women's Health Book Collective. 2005. *Our Bodies, Ourselves: A New Edition for a New Era.* New York: Simon and Schuster.

———. 1984. *The New Our Bodies, Ourselves: A Book by and for Women*. New York: Simon and Schuster.

———. 1976. *Our Bodies, Ourselves: A Book by and for Women*. New York: Simon and Schuster.

Casper, Monica, and Adele Clarke. 1998. "Making the Pap Smear into the Right Tool for the Job: Cervical Cancer Screening in the United States, c. 1940–1995." *Social Studies of Science* 28, no. 2/3: 255–90.

Charry, Tamar. 1997. "The Offbeat Film Maker David Lynch Directs a Campaign for Unilever's Home Pregnancy Test." *New York Times* (June 26).

Clarke, Adele. 1998. *Disciplining Reproduction: Modernity, American Life, and "the Problem of Sex."* Chicago: University of Chicago Press.

Coggins, Chrissy. 2000. "God Doesn't Make Mistakes." *UNITE Notes* 18, no. 4: 7.

Cote-Arsenault, Denise, and Robin Marshall. 2000. "One Foot In—One Foot Out: Weathering the Storm of Pregnancy after Perinatal Loss." *Research in Nursing and Health* 23: 473–85.

Cunningham, F. Gary, and Norman F. Gant, Kenneth J. Leveno, Larry C. Gilstrap III, John C. Hauth, and Katherine D. Wenstrom. 2001. *William's Obstetrics* (21st ed.). New York: McGraw-Hill.

Cunningham, G. C. 1976. "Letter: California's Position on Pregnancy Testing." *American Journal of Public Health* 66: 191.

Davis-Floyd, Robbie, and Elizabeth Davis. 1996. "Intuition as Authoritative Knowledge in Midwifery and Home Birth." *Medical Anthropology Quarterly* 10, no. 2: 237–69.

Doshi, Mary L. 1986. "Accuracy of Consumer Performed In-Home Tests for Early Pregnancy Detection." *American Journal of Public Health* 76, no. 5: 512–14.

Drug Store News. 1991. The Family Planning Center: More Profit Potential (September 30). http://findarticles.com/p/articles/mi_m3374/is_n17_v13/ai_11330462.

Duden, Barbara. 1993. *Disembodying Women: Perspectives on Pregnancy and the Unborn*. Cambridge, Mass.: Harvard University Press.

Eglash, Ron, Jennifer L. Croissant, Giovanna Di Chiro, and Rayon Fouche. 2004. *Appropriating Technology: Vernacular Science and Social Power*. Minneapolis: University of Minnesota Press.

Entwistle, P. A. 1976. "Do-It-Yourself Pregnancy Tests: The Tip of the Iceberg." *American Journal of Public Health* 66: 1108–9.

Evans-Smith, Heather Gail. 2002. "Fear." *UNITE Notes* 21, no. 2: 3.

Fertility Plus. 2003. (January revision). www.fertilityplus.org/faq/hpt.html.

Fisher, Jennifer K. 2002. "This Is the Story of Our Daughter, Catalina Pearl Robertson." *Sharing* 11, no. 5: 6.

Gelis, Jacques. 1991. *History of Childbirth*. Translated by Rosemary Morris. Boston: Northeastern University Press.

Gurdon, John B., and Nick Hopwood. 2000. "The Introduction of *Xenopus laevis* into Developmental Biology: Of Empire, Pregnancy Testing and Ribosomal Genes." *International Journal of Developmental Biology* 44: 43–50.

Johnsen, Michael. 2003. "Digital Pregnancy Test Kit Launches May Trigger Category Sales Rebirth." *Drug Store News* (November 17).

Johnson, A. 1976. "Do-It-Yourself Pregnancy Testing: The Legal Perspective." *American Journal of Public Health* 66: 129–30.

Jordan, Bridgette. 1977. "The Self-Diagnosis of Early Pregnancy: An Investigation of Lay Competence." *Medical Anthropology* 1, no. 2: 1–38.

———. 1993 [1978]. *Birth in Four Cultures: A Cross-Cultural Investigation of Childbirth in Yucatan, Holland, Sweden and the United States* (4th ed.). Prospect Heights, Ill.: Waveland Press.

Landsman, Gail Heidi. 2009. *Reconstructing Motherhood and Disability in the Age of "Perfect" Babies*. New York: Routledge.

Layne, Linda L. 2003. *Motherhood Lost: A Feminist Account of Pregnancy Loss in America*. New York: Routledge.

———. 2006a. "Unintended Consequences of New Reproductive and Information Technologies on the Experience of Pregnancy Loss." In Sue Rosser, Mary Frank Fox, and Deborah Johnson, eds., *Women, Gender and Technology*, 122–56. Urbana: University of Illinois Press.

———. 2006b. "'A Women's Health Model for Pregnancy Loss': A Call for a New Standard of Care." *Feminist Studies* 32, no. 3: 573–600.

———. 2007. "Designing a Woman-Centered Health Care Approach to Pregnancy Loss: Lessons from Feminist Models of Childbirth." In Marcia Inhorn, ed., *Reproductive Disruptions: Gender, Technology*, 79–97. Oxford: Berghahn Books.

Leavitt, Sarah A. 2005. "A Thin Blue Line." http://echo.gmu.edu/old/nih/responses.php. (Accessed May 20, 2005).

Leland, John. 2005. "U.S. Weighs Whether to Open an Era of Rapid H.I.V. Detection in the Home." *New York Times* (November 5): A11.

Maines, Rachel P. 2003. "Situated Technology: Camouflage." In Nina E. Lerman, Ruth Oldenziel, and Arwen P. Mohun, eds., *Gender and Technology: A Reader*, 98–122. Baltimore: Johns Hopkins University Press.

Martin, Joyce A., Brady E. Hamilton, Paul D. Sutton, Stephanie J. Ventura, Fay Menacker, Sharon Kimeyer, and Martha L. Munson. 2007. "Births: Final Data for 2005." *National Vital Statistics Reports* 56, no. 6. U.S. Department of Health and Human Services.

Nelkin, Dorothy, and Laurence Tancredi. 1994. *Dangerous Diagnostics: The Social Power of Biological Information*. Chicago: University of Chicago Press.

Oakley, Deborah. 1976. "Letter: On a Do-It-Yourself Pregnancy Test." *American Journal of Public Health* 66, no. 5: 502.

Oudshoorn, Nelly. 1994. *Beyond the Natural Body*. London: Routledge.

———. 2003. *The Male Pill: A Biography of a Technology in the Making*. Durham, N.C.: Duke University Press.

———, and Trevor Pinch, eds. 2005. "Introduction: How Users and Non-Users Matter." In *How Users Matter: The Co-Construction of Users and Technologies*, 1–28. Cambridge, Mass.: MIT Press.

Pfizer. www.pfizer.com/do/counter/womens/mn_wpt.html. (Accessed May 22, 2005).

Planned Parenthood. www.plannedparenthood.org/ABORTION. (Accessed May 25, 2007.)

Shew, M. L., W. L. Hellerstedt, R. E. Sieving, A. E. Smith, and R. M. Fee. 2000. "Prevalence of Home Pregnancy Testing among Adolescents." *American Journal of Public Health* 90: 974–76.

Stim, E. M. 1976. "Do-It-Yourself Pregnancy Testing: The Medical Perspective." *American Journal of Public Health* 66: 130–31.

Swain, Heather. 2004. *Luscious Lemon*. New York: Downtown Press.

Taylor, Janelle S. 2000a. "An All-Consuming Experience: Obstetrical Ultrasound and the Commodification of Pregnancy." In Paul Brodwin, ed., *Biotechnology and Culture: Bodies, Anxieties, Ethics*, 147–70. Bloomington: Indiana University Press.

———. 2000b. "Of Sonograms and Baby Prams: Prenatal Diagnosis, Pregnancy, and Consumption." *Feminist Studies* 26, no. 2: 391–418.

———, Linda Layne, and Danielle Wozniak, eds. 2004. *Consuming Motherhood*. New Brunswick, N.J.: Rutgers University Press.

Thomas, Sarah, and Charlotte Ellertson. 2000. "Nuisance or Natural and Healthy: Should Monthly Menstruation be Optional for Women?" *The Lancet* 355, no. 9207: 922–24.

U.S. Department of Health and Human Services. 1999. *Child Health USA 1999*. Washington, D.C.: U.S. Government Printing Office.

Vaitukaitis, Judith L., Glenn D. Braunstein, and G. T. Ross. 1972. "A Radioimmunoassay Which Specifically Measures Human Chorionic Gonadotropin in the Presence of Human Luteinizing Hormone." *American Journal of Obstetrics and Gynecology* 113: 751–58.

Wilcox, Allen J., Clarice R. Weinberg, John F. O'Connor, Donna D. Baird, John P. Schlatterer, Robert E. Canfield, E. Glenn Armstrong, and Bruce C. Nisula. 1988. "Incidence of Early Loss of Pregnancy." *New England Journal of Medicine* 319, no. 4: 189–94.

———, Donna Day Baird, David Dunson, Ruth McChesney, Clarice R. Weinberg. 2001. "Natural Limits of Pregnancy Testing in Relation to the Expected Menstrual Period." *Journal of the American Medical Association* 286: 1759–61.

Winner, Langdon. 1986. *The Whale and the Reactor: A Search for Limits in an Age of High Technology*. Chicago: University of Chicago Press.

Winston, Carla A., and Kathryn S. Oths. 2000. "Seeking Early Care: The Role of Prenatal Care Advocates." *Medical Anthropology Quarterly* 14, no. 2: 127–37.

Yankauer, Alfred, M.D. 1976. "Editorial: Do-It-Yourself Pregnancy Testing . . . Envoi." *American Journal of Public Health* 66: 131–32.

4 Breast Pumps

A Feminist Technology, or (yet) "More Work for Mother"?

KATE BOYER AND MAIA BOSWELL-PENC

INTRODUCTION

THIS CHAPTER QUERIES THE extent to which the breast pump can be considered a feminist technology. We approach this question through an analysis of how breast pumps affect women's mobility after childbirth and experiences of trying to combine milk expression with wage work in the United States. We explore the ways in which this technology can be considered liberatory and/or empowering and, if so, for whom.[1] On the one hand, breast pumps can be seen as one in a long line of technologies designed to mediate and manage women's bodies—from mammogram machines and IUDs to tampons, home pregnancy tests, menstrual-suppression technologies, and subcutaneous birth control devices, as explored in this volume. On the other hand, this artifact shares characteristics with devices intended to deliver more temporal and spatial freedom to their users, such as cell phones, laptops, and personal data assistants.

On average, women in the United States are allowed less maternity leave than in nearly any country on earth (Seager 1997). In the United States it is estimated that about seven in ten mothers with children under three years old work full time, while between one-third to one-half of working mothers return to work within three months after childbirth[2] and the remainder return within six months (Boston Women's Health Book Collective 2008, 273). When viewed in the context of the wage workplace we ask: Does the breast pump provide a means of "pushing back" on the work-life balance—and challenge traditional gendered public/ private divides—by bringing to work an activity traditionally associated with the private space of the home? Or, despite its promise, does the breast pump unwittingly create "more work for mother," as Ruth Schwartz Cowan (1983) argues that so many so-called labor-saving technologies have done? After providing a short background to the cultural context in which the breast pump emerged, we examine the experiences of women who have tried to combine lactation with wage work. We argue that although breast pumps can be considered a liberatory technology in that they expand women's choice in terms of feeding their infants, this potential is constrained by attitudes about pumping (and nursing more generally) as well as workplace design. Finally, we suggest a few possible pathways by which the breast pump's liberatory potential could be augmented.

We would like to clarify that we count women's unpaid labor in the space of the home as "work," and we have written elsewhere about the high cost of denying the social importance of care work (Boyer 2003). Yet this research focuses on pumping in workplaces outside the home both because most mothers in the United States return to work before their child is ready to eat solid food, and because pumping at work poses special problems and challenges that pumping at home does not.

Our research is informed by scholarship in feminist science and technology studies on the ways in which technology shapes and is shaped by social relations (Bray 1997; Cockburn and Ormrod 1993; Gorenstein 2000; Mackenzie and Wajcman 1985; Wajcman 1991, 2004; Webster 1996). This work has highlighted the ways gender, ethnicity, class, race, and other factors can be "built-in" to technology at the design phase (Cockburn and Ormrod 1993; Nelson, Tu, and Headlam Hines 2001). This scholarship has also challenged claims advanced by some as to the capacity of artifacts to change structures of power and inequality on their own (Cockburn 2004; Cowan 1983). Our analysis draws on Boswell-Penc's research on breast milk contamination and Boyer's research on gender, technology, and wage work. Boswell-Penc is the mother of two children and while breast-feeding was an avid pump user. Boyer has one child and has also experienced the breast pump as a user. We are both interested in this technology's potential to expand women's mobility, as well as its potential to redefine understandings of "appropriate" behavior and ways of being in the workplace. We base our analysis on twelve one-on-one interviews conducted between the spring and fall of 2005 with friends and acquaintances who had used breast pumps, employing a snowball methodology to constitute our interview pool. We also draw upon primary print and Web material on breast pump manufacturers, breast pump users, and breast pump advertisements, as well as secondary academic literature relating to the broader politics of infant feeding.

BACKGROUND

Breast pumps emerged out of two cultural trends occurring during roughly the same time period. The first of these was the growing recognition of breast milk's nutritional superiority to formula in the 1960s and 1970s, pioneered by lactation advocacy groups such as La Leche League. The second was the rise in the number of U.S. women returning to the wage workplace sooner after childbirth since the mid-1970s as a result of wage compression, the decline of fordism, and economic restructuring; together with women's advances into better-paying jobs higher up in managerial hierarchies (Hayghe 1986). Although the economic factors requiring greater numbers of women to return to the workplace sooner after childbirth affected women (and men) across race, ethnic, and economic lines, the question of how to proceed with infant nutrition in this context has produced a range of different responses.

Starting in the 1930s, the use of baby formula became widespread both domestically and as an export to the global south by companies such as Nestlé. As Boswell-Penc (2006), Hausman (2003), and others have argued, formula was advertised as being easier and less cumbersome than breast-feeding, and (erroneously) as being nutritionally superior to breast milk. Together with the rise of agrochemistry, bioengineering, and the shift toward more highly processed foods, formula fit within a modernist approach to health and nutrition in which more highly engineered products and practices were viewed as superior to their lower-tech alternatives (Apple 1987; Palmer 1988; Shapiro 2004). Formula remains big business, and the United States has one of the lowest rates of breast-feeding of any country in the developed world. According to a 2009 study by the U.S. Centers for Disease Control, only 35 percent of newborns in the United States are still being breast-fed at six months of age, and by one year only 16 percent receive any breast milk (www.breastfeed.com). Meanwhile, formula is routinely given out to new mothers in the maternity wards of many hospitals. Rates of breast-feeding among women who work outside the home are lower than for those who do not (Blum 1993, 296).

CONSIDERING BREAST PUMPS AS A FEMINIST TECHNOLOGY

Despite formula's dominance, scientific evidence now clearly shows breast milk to be nutritionally superior to formula, even when taking into account environmental toxins breast milk may contain (Boswell-Penc 2006; Hausman 2003). Infants who receive breast milk typically have fewer ear infections and respiratory infections, and fewer problems with diarrhea, among other immunological benefits (www.breastfeed.com). Thus, breast pumps provide a direct benefit for babies in terms of better health, as well as an indirect benefit for their mothers and/or other caregivers because fewer illnesses for the child also means fewer sick days and thus fewer unwanted disruptions to the caregiver's schedule. In addition, breast-feeding—even with a pump—is much less expensive than formula when considered over time. Although one year's worth of baby formula costs more than

$2,000, a new pump costs between $30 and $300 (www.breastfeedingonline.com). Initiatives such as the Boston-based "Pumps for Peanuts" project has sought to subsidize the cost of breast pumps for low-income groups, and it is also possible to lease pumps, for which prices vary.[3]

We would like to raise three points relating to the economics of pumping. First, in spite of the "value for money" of pumping versus formula when amortized over time, formula is cheaper than a pump (especially a high-quality one) at any single point of purchase. Second, although the U.S. Food and Drug Administration, La Leche League, and certain manufacturers discourage sharing breast pumps, our interviews as well as Web sites such as e-bay and Craigslist speak to the fact that a significant number of breast pumps continue to circulate beyond their first user (for example, a search for "breast pumps" on February 6, 2008, on e-bay produced a list of 889 items). Our third point is that depending on how much milk a woman wants to express, a pump may not be necessary. La Leche League's landmark book *The Womanly Art of Breastfeeding* (first published in 1963) explains how to express breast milk by hand, as does the *Our Bodies, Ourselves Pregnancy and Birth Book* (Boston Women's Health Book Collective 2008, 270–71).[4] That manual milk expression is not more common in the United States may be because of a range of factors, including lack of awareness about this technique, difficulty in successfully executing it, finding this technique too slow and/or laborious (or the perception that it would be), discomfort with the prospect of manually expressing one's own milk, the perception that a machine "will do it better," or simply the fact that there isn't any money to be made in it. Although none of the women in our study was able to express enough milk quickly enough by hand to make this a viable way to combine lactation with full-time wage work, we support hand-expression as an alternative method, and would support initiatives whereby this "no-tech," "no-purchase-necessary" method might gain wider recognition and acceptance.

Although we were not able to find data on breast-pumping rates in the United States, one study of 346 mothers conducted in early 2000 found that 77 percent of all mothers who breast-fed had used a pump at some point (Geraghty et al. 2005). Despite the health benefits of nursing, the decision to breast-feed, knowledge about the immunological benefits of breast milk, and awareness of breast-pumping alternatives are all structured by cultural background, social networks, and education. As of 2003, rates of breast-feeding in the United States were highest among college-educated white and Latina women over age thirty. Rates were lowest among mothers under age twenty, African American women, women who had not completed high school, and those living in the Southeast (*Child Health USA* 2003) (see page 123 for data on breast-feeding rates in the United States by race).

So to some extent, at present, breast pumps are a classed and racialized technology in that their users are not representative of women with young children as a group, but rather are a whiter and better-educated subset of that group.[5]

It should also be noted that breast feeding rates and demographics have been changing over time; in particular, between 1990 and 2001, the "race gap" for in-hospital breast-feeding initiation between African American women and other groups has begun to narrow (*Child Health USA*, 2003). These changes are likely a result of some combination of public health campaigns designed to raise awareness about the immunological benefits of breast-feeding combined with changing hospital protocols. However, although most U.S. women now do try breast-feeding in the hospital, rates thereafter (including the period during which women might begin to pump) begin to decline across all racial groups. Moreover, as of 2004 only 12 percent of U.S. women were breast-feeding exclusively for the first six months, as recommended by the World Health Organization.[6]

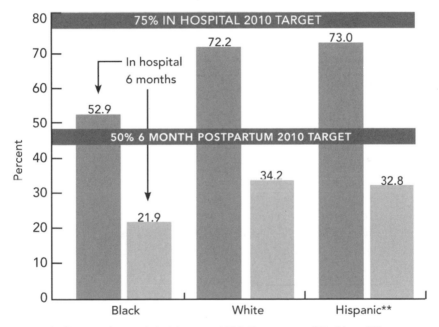

Breast-feeding rates by race/ethnicity: 2001.* U.S. Department of Health and Human Services, Health Resources and Services Administration, Maternal and Child Health Bureau. *Child Health USA* 2003. Available at: http://mchb.hrsa.gov/chusa03/pages/status.htm#breastfeeding.
*includes exclusive and supplemented breast-feeding
**may be of any race

Nevertheless, the greater availability of breast pumps since the 1990s has expanded the field of choices in the realm of infant nutrition, and may be working to help raise breast-feeding rates. Choice is important because, as Bronet and Layne note (this volume), women are not a homogeneous group and will have different needs and wishes. While recognizing that the benefits of this technology are not distributed evenly across all segments of society, we nevertheless suggest that this technology can be considered liberatory in that it gives nursing women more spatial freedom and allows more physical separation between mother and child, because even women who are committed to breast-feeding can sometimes feel "engulfed" by their baby. It also allows breast-feeding women to get someone else to take over one or more of the night or early-morning feeds. Breast pumps even provide a benefit for women who choose not to breast-feed as a means of providing relief from breast engorgement after one's milk comes in. They are also sometimes used by women who want to nurse an adopted infant because pumping (and nursing) can stimulate milk production through nipple stimulation (Boston Women's Health Book Collective 2005, 482). We suggest that in these respects at least, breast pumps function as a feminist technology.

BREAST PUMPS, MOBILITY, AND VISIBILITY

The modern breast pump was designed in 1956 by the Swedish/Swiss team of Einar Egnell and Olle Larsson, who went on to found the company Medela, which dominates the pump market globally (Bazelon 2006). Like tampons (Vostral, this volume), breast pumps were first designed for medical use, specifically for use in hospitals in cases where babies were too sick to breast-feed. In 1996, Medela revolutionized the breast pump market by releasing a breast pump designed for portability and personal use. Whereas these early models weighed about sixteen pounds (Bazelon 2006), contemporary portable models are considerably lighter, with common double-electric models (which are the heaviest models) weighing between nine and eleven pounds. Breast pumps now come in a wide range of designs: manual, electric, and battery-powered, as well as versions that are designed to pump either one breast or both at once (see page 125 for an image of a double-electric breast pump). Because breast pumps are competing with formula, their advertisements stress their liberatory potential by emphasizing ease of use and portability (page 125). Pump manufacturers also highlight the unobtrusiveness of their wares, as seen, for example, in Philips Avent's Isis "back-to-work" model, which is advertised as being especially quiet and discreet.

Starting in 2003, certain manufacturers began to offer "hands-free" models, one of which has been advertised in baby magazines through an image of a woman

Stealth pump: Boyer modeling a "briefcase"-style breast pump. Photo by Maia Boswell-Penc, Albany, New York, 2006.

Double-electric "briefcase"-style breast pump. Photo by Maia Boswell-Penc, Albany, New York, 2006.

supposedly pumping breast milk while pushing a child on a swing. This image could be read as a message that pumping can take place effortlessly amid one's day-to-day activities or, alternatively, as evidence of a culture that values and expects multitasking, in which "just nursing" is not enough.

Advertisements such as these transmit two conflicting messages. On the one hand, they suggest that this artifact will deliver greater autonomy and spatial freedom—desirable qualities. At the same time, these ads also reflect the anxiety that others might find out that one is pumping. Despite the fact that breast-feeding is starting to be protected by law,[7] in many places it continues to be perceived as scandalous and/or socially unacceptable. Like other female discharges, breast milk is often viewed in the popular imagination with an admixture of suspicion, fear, and disgust (Douglas 1966). Thus, although nursing may be considered the "natural" choice, the act itself is to be hidden (even if hidden in plain sight, as in the case of the mother pushing her child on a swing in a public playground). This echoes the way in which tampons work to hide the act of menstruation from the human eye (Vostral, this volume).

This raises a broader question about the politics of breast pump design, particularly in terms of who gets to set design objectives, what metrics are used to evaluate what constitutes a successful end-product, and how both of these have changed over time. Originally designed by (male) scientists and engineers, early product evaluations privileged science-based metrics such as prolactin and oxcytocin yields over other possible factors (Zinamen et al. 1992). But the breast pump underwent an interesting transformation during the 1990s as it shifted from a primarily hospital-based technology to a product on retail shelves. Whereas there are not clear ways to give feedback on the technologies one encounters as a patient (think, for example, of mammogram machines and speculae), a raft of Web sites now offer customer-generated product reviews of the different varieties of breast pumps on the market. Those in the market for breast pumps are very likely to have Internet access and very likely to already be in the habit of looking to Web sites to get advice, guidance, and support on topics relating to pregnancy and childbirth. Today's user may factor in considerations not only of efficacy but also comfort, noise level, or product weight (which bears upon portability), and are likely to look to feedback from other users (in both the virtual and nonvirtual world) in deciding which model to buy or lease.

The repositioning of users from patients to consumers with high levels of choice and means of communicating with each other raises an interesting potential for users to play an expanded role in product design and evaluation in the future. As Bronet and Layne note in this volume, the design field is still male-dominated in the United States. Although breast pump design and redesign has been led primarily by men with formal expertise in the fields of science and engineering, one now sees the potential for this process to accord more value to different kinds of voices and knowledges; in particular, the lay expertise of users. And indeed, one can find evidence that this is beginning to occur; for example, on the homepage for Philips Avent, a leading breast pump manufacturer based in the United Kingdom, a link invites visitors to "get involved with Philips Avent product development."[8] We welcome this shift, and suggest that the inclusion of a range of different kinds of users (and potential users) in the design process to schedule work breaks that coincide with let-down times."[14] Comments such as these suggest that it is up to individual employees to convince one's employer to allow them to pump at work. Most workplaces do not provide additional breaks to lactating women; thus, as Haidee Allerton points out in her article entitled "Coffee . . . Uh, Milk . . . Break," many women report having to accomplish their pumping during lunch or coffee breaks, around the edges of their workday (Allerton 1997).

Presumably, savvy employers in highly remunerative sectors of the labor market who value their female employees' time and want them back at work as quickly as possible after childbirth should provide time and space to allow their employees to blend nursing or pumping with wage work. And indeed, as feminist geographers Mona Domosh and Joni Seager note, one of the leaders in workplace lactation practices has been financial company J. P. Morgan Chase, which provides a lactation room with built-in pumps on their trading floor (Domosh and Seager 2001). Of course, most women do not work at J. P. Morgan Chase. And as a corollary to the gusto with which some high-wage employers have embraced workplace lactation, we suggest that finding time and space to pump is especially difficult in low-wage jobs. Space costs, so women who work in fast-food restaurants, coffee shops, or mall stores are very unlikely to have access to the kind of "extra spaces" such as dedicated lactation rooms, examination rooms, or empty conference rooms that middle-class women might use. Thus, whether one chooses to continue nursing after returning to work is not only a question of cultural preferences or the cost of a breast pump; it is also an issue of workplace design and whether one works for an employer who will make space for this activity.

CONCLUSION

Breast pumps, then, function as part of a broader sociotechnical system that includes workplace design and the social politics of actually pumping. Although we argue that the breast pump is emancipatory in that it expands mobility for some women, we also suggest that current social factors in the United States constrain this technology's liberatory potential. We are concerned that the "goods" of breast pumping are distributed unevenly to women employed in workplaces that will make room for pumping, and we postulate that one reason low-income women do not continue nursing as long as their middle-class counterparts is because they are less likely to work in such environments. At the same time, we are also concerned that by providing a personalized, technical fix to the question of workplace lactation, breast pumps could inadvertently remove the incentive for employers

represents a way to deliver even more fully on the breast pump's promise as a feminist technology.

BODIES OUT OF PLACE

We have argued that breast pumps can be considered a feminist technology in that they expand both mobility and choice in the field of infant nutrition. We now turn to what happens when users attempt to deploy this technology in real-world situations, particularly in the wage workplace. Although breast pumps can enable lactating women to return to work, actually pumping at one's place of employment can be difficult. At the most practical level, pumping at work is not a guaranteed right in every state, and women have been fired for it (Blum 1993, 291).[9] New York Congresswoman Carolyn Maloney noted as a reason for introducing the Pregnancy Discrimination Act Amendments of 2000, designed to protect woman's right to pump at work: "Women who choose to breastfeed have no choice about pumping milk during the day; they simply must express milk regularly." Maloney continued: "When women have stood their ground and told unrelenting bosses that, like it or not, they need [break time] for pumping, these same women have had their pay and benefits docked, and even lost their jobs."[10]

Indeed, even the breast-feeding advocacy community has been ambivalent about using breast pumps as a means to return to work. As Hausman has noted, there is a tendency within this community to view the breast pump as a technology that diminishes the emotional and psychological aspects of nursing (Hausman 2003), and historically this community's stance has been that women with young children should stay at home and nurse as long as they can (Bobel 2001). This position might be interpreted as only qualified support for women seeking to combine work and nursing, or possibly an anti-technology, anticonsumerist preference for manual expression, as discussed earlier.[11]

Even where pumping in the workplace is legal, problems can remain both at the level of built form and cultural practice, in that most workplaces "design out" nearly all activities other than work itself; for example, worksites in which one cannot eat or sit down (such as most stores) and in which bathrooms are too few (such as outside and/or male-dominated worksites) show how workplaces sometimes deny the physical needs of the body for employees of both sexes. Experiences of women seeking to pump and store milk at work, however, offer an illustration of how breast-pumping women—and their bodily products—are constructed as particularly "out of place" at work.

The lactating body in the workplace causes anxiety both because it draws attention to women's biological productivity and because it involves the purposeful excretion of a bodily fluid. As Boswell-Penc has argued, breast milk is sometimes viewed with suspicion as a potentially contaminated and contaminating substance (Boswell-Penc 2006). This builds on Mary Douglas's argument in *Purity and Danger,* which states that across many cultures, bodily fluids have been understood as dirt or pollution because they traverse what is usually the firm boundary of the body, thereby functioning as a symbol of danger, disorder, and power. Transgressions of the body boundary are fraught with ritual and particular social codes and, in many cultures, the bodily fluids of one sex are thought to pose a particular threat to the other (Douglas 1966, 120–21). Yet along side this disparaging view of breast milk stands an opposite interpretation: that of breast milk as "liquid gold," thus considered for its unique immunological and hormonal benefits that science has yet to find a way to replicate. In this respect, breast milk perhaps bears something in common with the way Waldby and Mitchell argue other "mobile" bio-objects are now viewed, as alternately precious or threatening depending on the circumstances (Waldby and Mitchell 2006).

We can find contemporary references to the fear breast milk can illicit in the 2003 case of an Albany, New York, woman who was fired for storing her breast milk in a communal refrigerator at her work.[12] This kind of anxiety also echoes old narratives of fear about women's entrance into the white-collar workplace in the early twentieth century, which produced impressive efforts to physically separate men and women employees and provide for women's "special needs" (a code word for menstruation) such as by providing spacious anterooms to bathrooms complete with sofas that could be used for rests (Boyer 1998). Such efforts can be interpreted variously. In one reading, such interventions reflect an arguably regressive desire to contain or quarantine women's bodies and their effluvia from the rest of the workplace. However, if we instead view these early spatial interventions as being of a piece with the provision of lactation rooms in today's offices, an alternative reading is that such accommodations instead constitute reasonable—indeed progressive—efforts to acknowledge and accommodate women's particular biological realities within the wage workplace.

As opposed to technologies that hide uniquely female bodily functions (such as tampons) that, as Vostral argues, are designed to *un*mark women's bodies (Vostral, this volume), the breast pump calls attention to a uniquely female bodily function, sometimes in a fairly dramatic way. Electric pumps, which were favored by nearly all of our interviewees for their efficiency, are also quite loud, and this was a source of anxiety for lactating women who chose to pump at work. As noted, in both pump advertisements and Web sites devoted to breast pumping, the noise level of breast pump motors is identified as problematic or embarrassing. For example, as Dilys Wynn, one nursing mother interviewed in the article "Express

Yourself: How to Successfully Combine Breastfeeding and W[...] experience: "I didn't tell my manager or my co-workers tha[...] milk at work. . . . It was a male-dominated industry and it w[...] embarrassing." To mitigate her embarrassment, Wynn pumpe[...] medical office.[13] One woman in our sample used a manual [...] way to reduce noise, but said that she got tennis elbow as a r[...]

Pumping requires both time as well as a place that ideally i[...] sanitary. Pumping has to be arranged so as to correspond roug[...] of nursing. Because nursing is a case of supply meeting deman[...] need to express regularly by one means or another in order t[...] duce milk. Our research suggests that this amounts to breaks [...] twenty-five minutes every few hours. In pumping advertisem[...] user groups, the need for privacy is also identified as being essen[...] process. Our interviews reveal a variety of strategies to achiev[...] as the "appropriate" degree of isolation. None of the women w[...] access to dedicated lactation rooms. Some pumped in their c[...] rooms, and some found small unused rooms or closets. One i[...] a colleague who worked in state government who wound h[...] window curtain. Anxiety about discharging a bodily substanc[...] or drawing attention to one's breasts by having them manipul[...] chine within earshot of ones' colleagues, together with the desi[...] another layer of spatial and temporal discipline on top of thos[...] by one's job, may all serve as disincentives to pumping at work[...]

We further submit that the emphasis on concealment an[...] scribed in our interviews suggest that lactating women almost [...] embarrassed to be seen by others engaged in pumping than t[...] We did not ask about this directly, but it squares with our bro[...] We have each witnessed colleagues breast-feeding in semipub[...] ference rooms, classrooms, faculty meetings, stores, and resta[...] upstate New York, it is very difficult to imagine someone pum[...] aforementioned situations or spaces.

TIME, SPACE, AND CLASS

Further, bracketing any inconvenience or embarrassment pump[...] being able to engage in this activity in the first place depends [...] secure the requisite time and space. Here, the onus is on emplo[...] employers) to find a way to make pumping work. As Dilys Wy[...] to you to work out where to pump, where to chill and store yo[...]

to come up with more—and perhaps better—alternatives for women trying to combine work and nursing.

By way of concluding we briefly note an article that appeared in the March 2006 edition of the online magazine *Slate*, written by Emily Bazelon. The article is entitled: "Milk Me: Is the Breast Pump the New BlackBerry?" (Bazelon 2006). By drawing the parallel, Bazelon was getting at the double-edged sword that is the BlackBerry and other such devices that allow us the temporal and spatial freedom to work outside of our allotted workplaces and work times. Although expanding one's temporal and spatial freedom sounds like a good thing, for many employees these devices have had a way of raising expectations about availability outside of work hours, as well as the amount of time spent on work-related activities outside the workplace (Wheeler, Aoyama, and Warf 2000, 31–41). Bazelon frames the comparison thus: "Like BlackBerrys, pumps give us freedom we otherwise wouldn't have in exchange for inviting us to go to lengths that we otherwise couldn't" (Bazelon 2006). Finally, Bazelon's article pushes us to ask: Despite the mobility-enhancing benefits breast pumps can provide, is there a risk that their existence could inadvertently foreclose other options for combining wage work and care work? Because (some) women now have the capacity to both work outside the home and continue to nurse, will they be pressured to do so? Could the existence of breast pumps be used against efforts to fight for longer maternity leaves (and paternity leaves) or the provision of on-site child care?

Some of these factors can be ameliorated by policy, as through legislation that would force employers to provide the time and space necessary to make pumping and/or nursing possible in *all* workplaces, as other bodily needs are met through the provision of bathrooms and, in some workplaces, first-aid stations.[15] Policy provides an important part of the solution because it has the power to reframe responsibility for a problem from the individual to the collective. Although at present most workplaces in the United States do not provide lactation rooms, incentives to businesses to adopt breast-feeding-friendly policies include savings in health care costs, prescription costs, and lost work hours, as scholars from the Tulane Xavier National Center of Excellence in Women's Health have observed.[16]

Other barriers can be solved through better design, such as through the development of pump models that are lighter and quieter (Perkins 1999), or through other design improvements that are as yet unimaginable. As noted, the shift toward including the perspectives of women in the process to achieve more user-centered designs could serve as an important pathway toward achieving this goal.

In addition to the material and policy changes that are required, we argue that for the breast pump to truly fulfill its potential as a feminist technology, it will require cultural changes as well.[17] We need to change the politics of banishment that accompany breast-feeding, and challenge narratives that encode woman's biological productivity as shameful. We need to expand the limits on what kind of bodies belong at work, and allow a broader range of living to occur there (especially as so much work has followed us home). Finally, rather than focusing single-mindedly on *this* way of blending nursing with other activities, we propose that it is important to remain open to finding *even better* ways, as well as remaining mindful of preserving choice, because this is obviously not a domain in which one solution will fit all. These transformations will require legislative change and changes to the material environment as well as cultural change.

NOTES

1. This paper was developed out of our article: "Expressing Anxiety? Breast Pump Usage in American Wage Workplaces," Gender, Place and Culture 14, no. 5 (2007): 551–67.

2. http://www.tulane.edu/~tuxcoe/NewWebsite/com_womens_health/pdf/workingmomsbreastfeeding.pdf. (Accessed March 23, 2008.)

3. We were not able to find data on how common breast pump leasing is compared with buying.

4. "Hand Expression," *New Beginnings* 13, no. 2 (1996): 51–52. http://www.llli.org/NB/NBMarApr96p51.html. (Accessed February 7, 2007.) *Our Bodies, Ourselves* states that "while you are still in the hospital, a nurse or lactation specialist should coach you on how to express milk by hand" (Boston Women's Health Book Collective 2008, 271).

5. For more on breast-feeding as a luxury, see Blum 1993, 292.

6. Figures are from 2004's "Breastfeeding among U.S. Children Born 1999–2005, CDC National Immunization Survey, Centers for Disease Control, http://www.cdc.gov/breastfeeding/data/NIS%5Fdata/. (Accessed May 12, 2009.)

7. "Breastfeeding Legal," *Maclean's* 110, no. 30 (July 28, 1997): 29; "Breast-Feeding: A Civil Right," *New York Times* (May 20, 1994): A14, A26; "Florida Approves Public Breast-Feeding," *New York Times* (March 4, 1993): A8, A18.

8. http://www.avent.philips.com/en_GB/. (Accessed August 8, 2008.) Clicking the link directs visitors to a page where they can fill out contact details so they can become involved in future product development. More research is required to find out exactly how those who express interest in becoming involved are incorporated into the design process.

9. See also "Rep. Maloney Introduces Legislation to Give Women Legal Protection against Termination and Discrimination on the Job for Breastfeeding," http://maloney.house.gov/index.php?option=com_content&task= view&id=831&Itemid=61. (Accessed April 15, 2006.)

10. See also http://maloney.house.gov/index.php?option=com_content&task=view&id=831&Itemid=61. (Accessed April 15, 2006.)

11. Bobel suggests that La Leche League's basic stance is that staying at home to nurse is preferable to returning to work, even with a pump (2001:139).

12. In *Kathleen Landor-St. Gelais v. Albany International Corporation,* July 31, 2003, the plaintiff sought to pump milk at work (in a bathroom stall) and to store it in a communal refrigerator; her suit was dismissed on several grounds (State of New York Supreme Court, Appellate Division, Third Judicial Department).

13. Froud, Helen, "Express Yourself! How to Successfully Combine Breastfeeding and Work" http://breastfeed.com/resources/articles/bfeedandwork.htm. (Accessed April 15, 2006.)

14. Froud, Helen "Express Yourself! How to Successfully Combine Breastfeeding and Work" http://breastfeed.com/resources/articles/bfeedandwork.htm. (Accessed April 15, 2006.)

15. Some of the most active work in the realm of nursing and breast-pumping legislation has been done by Carolyn B. Maloney from the Eleventh District of New York, Manhattan and Queens. See, for example, the New Mother's Breastfeeding Promotion Act, introduced in 2003, which would protect women from discrimination in the workplace for pumping or breast-feeding. This act would also give tax incentives to employers to set up lactation rooms. "Fighting for New Mothers," http://maloney.house.gov/index.php?option=com_content&task=view&id=419&Itemid=61. (Accessed May 12, 2009.)

16. http://www.tulane.edu/~tuxcoe/NewWebsite/com_womens_health/pdf/workingmomsbreastfeeding.pdf. (Accessed March 23, 2008.)

17. For an analysis of different methods of "fixing" issues in science and technology that challenge the cultural mainstream, see Layne 2000.

References

Allerton, Haidee. 1997. "Coffee . . . Uh, Milk . . . Break." *Journal of Training and Development* (November): 10.

American Academy of Pediatrics. 2005. "Policy Statement: Breastfeeding and the Use of Human Milk." *Pediatrics* 115, no. 2: 496–506.

Apple, Rima. 1987. *Mothers and Medicine: A Social History of Infant Feeding, 1890–1950.* Madison: University of Wisconsin Press.

Avent. http://www.avent.philips.com/en_GB/. (Accessed August 8, 2008.)

Bazelon, Emily. 2006. "Milk Me: Is the Breast Pump the New BlackBerry?" *Slate* (March 27). http://www.slate.com/id/2138639/?nav=navoa. (Accessed May 12, 2009.)

Blum, Linda. 1993. "Mothers, Babies, and Breastfeeding in Late Capitalist America: The Shifting Contexts of Feminist Theory." *Feminist Studies* 19, no. 2: 291–311.

Bobel, Christina. 2001. "Bounded Liberation: A Focused Study of La Leche League International." *Gender and Society* 15, no. 1: 130–51.

Boston Women's Health Book Collective. 2005. *Our Bodies, Ourselves: A New Edition for a New Era.* New York: Simon and Schuster.

———. 2008. *Our Bodies, Ourselves: Pregnancy and Childbirth.* New York: Simon and Schuster.

Boswell-Penc, Maia. 2006. *Tainted Milk: Breastmilk, Feminisms, and the Politics of Environmental Degradation.* Albany, N.Y.: SUNY University Press.

Boyer, Kate. 1998. "Place and the Politics of Virtue: Clerical Work, Corporate Anxiety, and Changing Meanings of Public Womanhood in Early Twentieth-Century Montreal." *Gender, Place and Culture: A Journal of Feminist Geography* 5, no. 3: 261–76.

———. 2003. "At Work, At Home? New Geographies of Work and Caregiving under Welfare Reform in the U.S." *Space and Polity* 7, no. 1: 75–86.

Bray, Francisca. 1997. *Technology and Gender: Fabrics of Power in Late Imperial China.* Berkeley: University of California Press.

Centers for Disease Control. 2009. "Breastfeeding among U.S. Children Born 1999–2006: CDC National Immunization Survey." http://www.cdc.gov/breastfeeding/data/NIS%5Fdata/. (Accessed May 12, 2009.)

Child Health USA. 2003. U.S. Department of Health and Human Services, Health Resources and Services Administration, Maternal and Child Health Bureau. http://mchb.hrsa.gov/chusa03/. (Accessed May 13, 2009.)

Cockburn, Cynthia. 2004. "Review of Wajcman, Judy, *TechnoFeminism*." *Women's Studies International Forum* 27: 605–6.

———, and Susan Ormrod. 1993. *Gender and Technology in the Making.* Thousand Oaks, Calif.: Sage.

Cowan, Ruth Schwartz. 1983. *More Work for Mother: The Ironies of Household Technology from the Open Hearth to the Microwave.* New York: Basic Books.

Curtis, Cindy. http://www.breastfeedingonline.com/pumps.html. (Accessed April 15, 2006.)

Domosh, Mona, and Joni Seager. 2001. *Putting Women in Place: Feminist Geographers Make Sense of the World.* New York: Guilford Press.

Douglas, Mary. 1966. *Purity and Danger: An Analysis of Concepts of Pollution and Taboo.* London: Routledge & Kegan Paul.

"Fighting for New Mothers." http://maloney.house.gov/index.php?option=com_content&task=view&id=419&Itemid=61. (Accessed May 12, 2009.)

Geraghty, Sheela, Jane Khoury, and Heidi Kalkwarf. 2005. "Human Milk Pumping Rates of Mothers of Singletons and Mothers of Multiples." *Journal of Human Lactation* 21, no. 4: 413–20.

Gorenstein, Shirley, ed. 2000. *Research in Science and Technology Studies: Gender and Work.* Stamford, Conn.: JAI Press.

"Hand Expression." *New Beginnings* 13, no. 2 (1996): 51–52. http://www.llli.org/NB/NBMarApr96p51.html. (Accessed May 12, 2009.)

Hausman, Bernice. 2003. *Mother's Milk: Breastfeeding Controversies in American Culture.* New York: Routledge.

Hayghe, Howard. 1986. "Rise in Mothers' Labor Force Activity Includes Those with Infants." *Monthly Labor Review* 109: 43–45.

http://www.breastfeed.com/news.htm. Web site of i-parenting media. (Accessed April 15, 2006.)

Layne, Linda. 2000. "The Cultural Fix: An Anthropological Contribution to Science and Technology Studies." *Science, Technology and Human Values* 25, no. 4: 492–519.

Mackenzie, Donald, and Judy Wajcman, eds. 1985. *The Social Shaping of Technology: How the Refrigerator Got Its Hum*. Milton Keynes and Philadelphia: Open University Press.

Medela. http://www.medela.com/graphics/pnsa_3womenftherW.jpg. (Accessed April 15, 2006.)

Nelson, Alondrea, Linh N. Tu, Thuy and Alicia Headlam Hines, eds. 2001. *Technicolor: Race, Technology, and Everyday Life*. New York: New York University Press.

Palmer, Gabrielle. 1988. *The Politics of Breastfeeding*. London: Pandora Press.

Perkins, Nancy. 1999. "Women Designers: Making Differences." In Joan Rothschild, ed., *Design and Feminism: Re-Visioning Spaces, Places, and Everyday Things*, 120–25. New Brunswick, N.J.: Rutgers University Press.

"Rep. Maloney Introduces Legislation to Give Women Legal Protection Against Termination and Discrimination on the Job for Breastfeeding." http://maloney.house.gov/index.php?option=com_content&task= view&id=831&Itemid=61. (Accessed April 15, 2006.)

Seager, Joni. 1997. *The State of Women in the World Atlas*. New York and London: Penguin.

Shapiro, Laura. 2004. *Something from the Oven: Reinventing Dinner in 1950s America*. New York: Viking.

Tulane Xavier National Center of Excellence in Women's Health. http://www.tulane.edu/~tuxcoe/NewWebsite/com_womens_health/pdf/workingmomsbreastfeeding.pdf. (Accessed August 20, 2008.)

United States House of Representatives. http://www.house.gov/maloney/issues/womenchildren/breastfeeding/worldwide.htm. (Accessed April 15, 2006.)

Vostral, Sharra. 2005. Hidden Disaster: Rely Tampons and the Story of Failed Design. Paper given at the Society for the Social Study of Science, Pasadena, California.

Wajcman, Judy. 1991. *Feminism Confronts Technology*. Cambridge, UK: Polity Press.

———. 2004. *TechnoFeminism*. Cambridge, UK: Polity Press.

Waldby, Catherine, and Robert Mitchell. 2006. *Tissue Economies: Blood, Organs and Cell Lines in Late Capitalism*. Durham, N.C.: Duke University Press.

Webster, Juliet. 1996. *Shaping Women's Work: Gender, Employment, and Information Technology*. New York: Longman Group.

Wheeler, James, Yuko Aoyama, and Barney Warf, eds. 2000. *Cities in the Telecommunications Age: The Fracturing of Geographies*. New York, Routledge.

Zinamen, Michael, Vergie Hughes, John Queenan, Miriam Labbok, and Barry Albertson. 1992. "Acute Prolactin and Oxytocin Responses and Milk Yield to Infant Suckling and Artificial Methods of Expression in Lactating Women." *Pediatrics* 89, no. 3: 437–40.

5 Tampons
Re-Scripting Technologies as Feminist

SHARRA L. VOSTRAL

"WELCOME THIS NEW DAY FOR WOMANHOOD" announced the first Tampax advertisement published in *The American Weekly* in 1936. This new day included technologically mediating the body with a mass-produced cotton tampon worn internally within the vagina to absorb menstrual flow. In this construction, women are newly liberated and handed physical freedoms as a result of the use of tampons. The advertisement depicts women riding horses and sunbathing at the beach, two activities difficult to accomplish with a bulky sanitary pad between the legs. In addition, because women were (and still are) socially required to conceal their periods, tampons aided women in accomplishing this social expectation. By virtue of their relationship to women's unique biological processes of menstruation, tampons are easily construed as a "feminine technology" (McGaw 2003, 15–16). But are they, or could they be, a feminist technology as well?

The exercise of engaging tampons as a feminist technology begets two strands of thought. The first is an examination of what constitutes a feminist technology, and how that might be defined. The second component requires a look into the design history of tampons and how they were marketed and sold, as well as how and when women used them. In addition, I argue that existing technologies such as the tampon can be redesigned and re-scripted to produce feminist technologies. I borrow the concept of scripting from Madeleine Akrich, who discusses how designers and inventors build into objects information about how the objects should be used (Akrich 1992, 209; Oudshoorn and Pinch 2003). If we can think of an object as text, then scripting transmits information about how to "read" it, or how to decode the object for use. An examination of the evolution of the tampon reveals that this re-scripting has produced feminist tampon technologies where previously none existed.

The topic of feminist technology is an important one because feminism has provided critical discourse to engage issues of gender and power in many arenas. Liberal feminism, radical feminism, socialist feminism, ecofeminism, and even scientific feminism are recognized mainstays that interrogate the place of women and gender in their respective philosophies. However, there is not a well-developed school of thought about feminist technology, although there is a growing body of work that encompasses gendered analyses of technologies. Adding technology to feminist agendas is necessary to ongoing struggles to achieve gender equity and equality because the technological often coproduces and maintains systems of power that affect women's lives. One of the perpetual problems of designing technologies is that the assumed user has been the "universal man." From cockpit design to the height of doorknobs, the universal male body reigns supreme, with both large-framed and small-framed humans, elderly adults and children, and people with disabilities left to the wayside. There has been a call for universalist design, which alleviates many of these problems, but it is a myth that a neuter design can accommodate one and all. For women, there are indeed pragmatic differences, such as biological ones like menstruation or even contextual ones such as safety shoulder belts in cars, that do require attention for optimal designs for women. Yet, when this difference is essentialized, and women become reduced to their reproductive functions, we are caught in the same old trap of difference being interpreted as unequal to and lesser than men. How might feminist designers wend their way through traps, while developing technologies that contribute to women's empowerment; that is, designing feminist technologies? A retrospective look at tampons, keeping in mind the intent of designers and the scripting and subsequent re-scripting of technologies, demonstrates how technological design has the potential to be imbued with feminist politics.

THE MENSTRUAL TAMPON: A FEMINIST INNOVATION?

During the early to mid-twentieth century in the United States, there were many companies that began to design menstrual hygiene products for women. This section details designs of the first manufactured tampons, in which engineers and inventors clearly had women in mind, and the technologies were scripted with very pragmatic intentions. Yet designing for women as consumers is not necessarily feminist, nor does it automatically yield feminist technology. In fact, the earliest tampons from the nineteenth century were not designed for women at all, but were medical devices to stop up gaping wounds. Gynecologists from that time period often administered tampons, soaked in mercury chloride, otherwise known as calomel, to treat vaginal discharge and brace the vaginal walls. Of the patents granted for tampons between 1890 and 1921, all referred to surgical or medical applications (Farrell-Beck and Klosterman Kidd 1996, 325). Additionally, tampons served contraceptive purposes when soaked with the appropriate compounds and inserted into the vagina. Because distribution and use of contraception were considered illegal (Brodie 1994, 224), tampons gained some notoriety because of the prevailing 1873 Comstock law regulating decency standards.

By the 1920s, part of displaying a modern identity for women meant managing menstruation with sanitary napkins, first advertised and sold immediately after World War I. Young women were more likely to purchase and use these disposable pads than their mothers. Although more convenient in some respects than reusable and washable rags, disposable pads were often bulkier and increased chafing of the thighs. The elastic belt and clasps were no panacea either because they shifted in the undergarments and were often quite awkward to wear. "What most of them wish (and as yet they have not had this wish granted)," noted researcher and efficiency expert Lillian Gilbreth in 1927, "is for a new product which will be completely invisible no matter how tight or thin their clothes are" (Gilbreth 1927, 70). Her market research indicated that women wanted manufactured menstrual hygiene products that would be imperceptible to the eye. Because menstruation proved to be a problem that technological innovation could address, inventors responded by developing and selling what they called an "internal sanitary napkin": the tampon. Inserted into the vagina and held in place by the vaginal walls, tampons proved to be an entirely different means to technologically manage menses. The tampon avoided the shortcomings and offered many benefits over sanitary pads; they better protected bedding and clothing from stains and obviated the need for washing menstrual rags. They concealed menstrual blood and menstrual scents because, unexposed to air, menstrual fluid does not decompose to release an odor. They left no outlines, required no belts or harnesses, and promoted mobility through bodily flexibility. On a pragmatic level, the technological fix offered an immediate physical freedom during menstruation previously unavailable by using rags, menstrual pads, and belts because the body remained unencumbered. Tampons also hid the evidence of blood, thereby protecting a menstruating woman from the judgments of men and women.

The menstrual tampon was "invented" independently by both a man and a woman at the same time. In 1927, Ives Marie Paul Jean Burill filed an application for a "catamenial appliance" that he claimed was a new means of absorbing menstrual flow. Because sanitary pads could aggravate lesions and irritate skin leading to possible infection, Burill argued that his appliance would stop blood without these kinds of effects. His design called for a core of touchwood to absorb most of the fluid, although it would be covered with cotton, cellulose, and gauze. A piece of colored silk string to tie off the gauze would be used to remove the plug once it was full (Patent 1,726,339 1929). Marie Huebsch also filed her patent for a hygienic device in 1927. She stated that hers was a device "directed to an improved simplified non-washable, pack-plug for absorbing the catamenial flow or other discharges from the vagina" (Patent 1,731,665 1929). The composition of the absorbent material could be cellulose derived from linen or cotton, wool, flax, or even wood fiber. The importance of her patent was not in materials but design. She proposed rolling the filler and coiling the external gauze to keep the filler tightly in place. By winding the gauze and extending it past the filler, it also served "to form a projecting handling end" in order to remove it from the vagina. Although both patents were granted, neither developed into successfully mass-produced items.

Two more patents were filed in 1931. One belonged to Frederick S. Richardson, who purported to be "the first to have provided any practicable form of cata-menial device intended and adapted for use inside the vagina" (Patent 1,932,383 1933). One of the differences with Richardson's design was in the wrapping of the string, and another was in offering different sizes. The plug itself was composed of cotton or cellulose, measured two and a half inches long by three-fourths of an inch in diameter. The moisture-pervious "slipper" of perforated cellophane wrapped the plug and held the material in place, and slid easily into the vaginal canal. The pull string though was threaded through the coiled layers of absorbent material and then laced out of the end of the plug. Because the products would be made in different sizes, women could choose the size to match their flow. Richardson managed to market his new product as Wix tampons, advertised as "sanitary protection," which alluded to a clean, disease-free item to prevent damage to clothing and women themselves. According to its ads, Wix eliminated "the embarrassment of protruding pads" and gave "complete, healthful protection *internally, invisibly*" ("Sanitary Protection" 1935, SM18, emphasis in original).

The other brand, whose patent was also filed in 1931, grew into the product named Tampax, manufactured by Tambrands, Inc. The inventor of Tampax tampons, Earle Cleveland Haas, an osteopath by training, admitted that he conceived of the design because he "just got tired of women wearing those damned old rags" (Bailey 1987, 5). He also pitied women during their menses. When asked how he became interested in designing a tampon, he replied, "Well, I suppose I thought of the poor women that got in that mess every month, you know. It was very disagreeable to them and I thought, well, surely there is a better way to take care of that" (*Kehm v. Proctor & Gamble*, 1173 [1982]). Haas's design prevented the "bloody mess." His cotton tampon was first stitched to-gether and then compressed. The "longitudinal stitching of the cotton" allowed for expansion lengthwise in which "the expanded core straightens within the vagina when drawn upon." The problem with other catamenial devices was that they "have been made to attach a tape or ribbon to a core of absorbent material but it was found that after the material expanded in the vagina it was exceedingly difficult to withdraw it owing to the restricted opening" (Patent 1,926,900 1933). With the threads stitched right into the cotton longitudinally, so that the cord extended beyond the bleached cotton, it was then compressed so that it could be inserted easily. It would then expand lengthwise inside the vagina as it absorbed fluid. In order to perfect the size of the cotton plug, he felt no compunction to measure the cervix or vagina because he had "seen so damn many of them" that he felt he got the gist. He believed Tampax would conform any woman's vagina, regardless of shape, size, or angle of the vaginal canal (*Kehm v. Proctor & Gamble*, 1179 [1982]).

Luckily for Haas, his wife, a nurse, helped him to hone the final form of the tampon by wearing various modifications throughout the developmental process. She also helped him by distributing test samples to other nurses. Although the plug was unique in that it expanded lengthwise, its real advantage came from his patented dispenser: a telescopic applicator made of disposable cardboard cylinders. Haas considered a reusable metal applicator—along the schematics of a speculum—but the convenience, and profitability, of a disposable applicator held sway. With the disposable cardboard, women would continually repurchase the same item, boosting profits. The cardboard applicator allowed women to insert the tampon without their fingertips ever touching their bodies, and in theory flush the applicator down the toilet after they finished. Because cardboard was water soluble, "they may be discarded into a toilet bowl where they will dissolve or soften so that they may be flushed away" (Patent 1,926,900 1933). Haas named the catamenial Tampax, for vaginal tampon packs, mimicking the hard final syllable of the sanitary napkin brand Kotex (Bailey 1987, 6).

In 1934 Haas sold his patent and company for $32,000 to Gertrude S. Tenderich, a Denver businesswoman and a German immigrant. As he put it, he "had a lot of trouble with this thing" and could not get anyone to buy the option for the company or the patent. In fact, by the time he found a buyer, he sold it outright with no possibility for royalties. By then, he admitted, "I didn't want anything to do with it. I was so sick of it. After a while you get turned off on those things, boy" (*Kehm v. Proctor & Gamble*, 1177–78 [1982]). Organizing Tampax Sales Corporation in 1934, Tenderich and her employees made small lots of tampons in a Denver warehouse. Personally delivering her sales pitch at drugstores, Tenderich tried to persuade pharmacists to purchase the tampons, but they often expressed little interest. Seeking to reach a wider market, Tenderich traveled to New York where she met prospective buyer Ellery Mann, who had been the president of Zonite Products Corporation, which manufactured douches. Mann was no stranger to feminine or personal hygiene products—he was also responsible for the sale of toothpaste, soap, and mouth spray (Bailey 1987, 8–10). He helped propel sales and laid the foundation for the company.

The origin story of Tampax and the manner in which it was designed does not read like a feminist tale, and Haas did not script feminism into Tampax. Based upon Haas's comments, he may have felt pity for women during their periods, but this attitude does not empower women. Sympathy might be detected in the desire to improve women's comfort, but his primary motivation was the pleasure of an inventive conundrum to solve, with the outcome being financial reward.

A far more frightening result of nonfeminist design was that of Rely tampons, manufactured by Proctor & Gamble during the late 1970s and early 1980s. Set on hobbling Tampax as the most popular tampon producer, Proctor & Gamble created and marketed an innovative tampon that held excellent potential for succeeding in this task. The tampon itself was different because it was composed of synthetic materials and not cotton like the others. Housed in a polyester-like pouch, foam cubes and carboxymethylcellulose chips absorbed menstrual fluids. This application of carboxymethylcellulose to tampons was novel, but it had been used as food thickener in puddings and pie. By industry standards it was considered safe because it was edible. It was the absorptive power of the chips that brought Rely a devoted following, and also caused its undoing. The use of these particular tampons brought on a set of symptoms in some women such as rash, fever, flu-like symptoms, and even shock. Physicians at the Centers for Disease Control scrambled to identify what they believed to be an emergent disease, and publicized the condition known as toxic shock syndrome (TSS). The tampons did not cause TSS per se, but were implicated as a catalyst in triggering the syndrome (Vostral 2008, 157–58).

Toxic shock syndrome was an unexpected consequence of using the new tampon, but there were two problems that made the situation worse. First, most of the ingredients for the tampon were "grandfathered" in as safe and did not require additional testing. Compounding this was the category of tampons as cosmetics until 1976, which offered the company a great degree of leniency and freedom in terms of safety testing. Afterward, they were categorized as medical devices with somewhat better oversight. As the women's health movement has highlighted, many products sold to women cause pain, suffering, and damage, and must be rigorously tested to ensure the safety of their use for all women. The second problem exacerbating the spread of TSS was not technological, but ethical. Leadership at Proctor & Gamble hid the severity of TSS from its consumers, and only removed Rely from store shelves when the Food and Drug Administration threatened a product recall. Furthermore, women's bodies were blamed for aggravating the disease, as if their dirty fluids and poor hygiene habits were the source of the problem. This behavior of the company to protect its reputation and wealth ultimately undermined women's health, and in 1982 the company was found negligent in a court trial in which it was sued for the wrongful death of Patricia Kehm, who died from TSS after wearing Rely tampons during her period.

On the other end of the spectrum, a tampon with seemingly feminist origins is o.b., because the brand claims that it was "created by a woman gynecologist." However, this commercial scripting of feminism was added during the early 1970s, nearly twenty years after the product's emergence on the market in Germany. This assertion in the advertising implies that the inventor was able to invent a tampon superior to other products because of her special training as a woman's physician and her essential nature as a biological female. Interestingly, the o.b. tampon probably did not originate with her, but with an engineer, Carl Hahn, backed by lawyer Heinz Mittag, in 1945. The woman gynecologist, Judith Esser (later Esser Mittag), served as a consultant, and probably even a major contributor, but she was not the sole inventor ("Geschichte des Tampons" n.d.). The entrepreneurs sold it in the German market in 1950 as o.b., an acronym for *ohne binde,* roughly translated as "without a pad." Johnson & Johnson later purchased the brand in the early 1970s at the apex of the Women's movement, and incorporated elements of feminism into its marketing ("Early Ad for o.b. Tampons" n.d.). Thanks in part to the success of the book *Our Bodies, Ourselves* and the women's health movement, the advertisements for o.b. emphasized the unique applicatorless design that liberated women were qualified to use because of their newfound comfort touching their vaginas. Because of the women's health movement, it was important to deploy rhetoric that a woman—a female gynecologist, no less—who presumably knew a thing or two about vaginas, designed the tampon. In this origin story, she was crafted as having unique insight into women's menstrual needs because she embodied the identity of "woman."

The o.b. tampon demonstrates a commercial claim to a feminist technology. The corporate story attempts to mythologize the lone female scientific hero, which misrepresents her role and therefore cannot be deployed to engender female empowerment. It also speaks to the perceived financial importance of appealing to liberal feminism in this consumerist way. Leaving aside, for a moment, the commercial pretensions of being feminist, how does this technology measure up? The unique feature of this design was the fact that it was applicatorless. As promoted by second-wave feminism and books like *Our Bodies, Ourselves,* gaining the right to touch one's vagina without shame was a significant achievement of women's liberation. The same small feature that required intimate knowledge of the vagina also allowed it to slip into a purse or pocket or be concealed in the hand. This has made it easier for women to hide their periods and to "pass" as nonmenstrual (Vostral 2008, 10). This is an important skill in a society that values clean, tidy bodies, but especially important in patriarchal societies that loathe women's menstrual blood. Adapting to such patriarchal values while pragmatic is also problematic. In the meantime, while radical change in thought is slow in coming, the applicatorless design of o.b. creates less waste, which is an advantage in terms of ecofeminism. All this highlights the need to continue the cultural work of validating menstruation, but also to design and promote feminist-technical practices for menstrual management.

ᕦᕤ

I see feminist innovations in tampon technology taking two forms. The first is to re-script those technologies already in use. In this section I introduce several examples of how familiar artifacts can be re-scripted as feminist and invested with new politics by users. The second approach that I see begins in the design process, when ideas are nascent and radical new conceptualizations can be envisioned with feminist design principles.

ECOFEMINIST INNOVATIONS

In response to a number of health dangers posed by tampons, as well as the ecofeminist movement, starting in the 1960s some women consciously chose to abandon the scripted artifacts of the dominant, commercial menstrual hygiene industry and sought out alternative methods to trap menstrual flow. Readily available at health food stores are washable receptacles, shaped in the form of a cup or diaphragm, and worn in the vagina for up to twelve hours. Called menstrual cups, these devices are reusable and washable, and appeal to women who want to diminish their reliance on disposable products, consistent with the tenets of ecofeminism, to minimize their environmental impact and pay respect to natural resources. An example of a menstrual ecofeminist is Emma, a student with whom I spoke at a small liberal arts college, who proudly revealed that the menstrual fluid she saved from her cup, when mixed with water, promoted the growth of her plants when she used it as a fertilizer. She thought her plants never looked so good. She also happily claimed the honor of being dubbed the "menstrual lady" after she decided to educate women on her campus about menstrual hygiene alternatives, such as her favorite, the Diva-Cup. She also earned the name by initiating sewing bees in which women learned to make their own flannel pads. The women chose their fabric design, anything from the ironic plain red to Scooby Doo, and she instructed them how to sew in an absorptive layer and where to put the Velcro. Emma said most women were amazed at how comfortable the pads were and appreciated the benefits of washing the pads instead of tossing them out with the trash (interview by author, May 26, 2005). Going beyond making pads, some women relinquish them altogether. There is a practice called free bleeding, in which women do not use anything to absorb their flow. Some practice this only while sleeping, and find that by propping up the pelvis, the vagina traps the blood, which can then be expelled in the toilet. Menstrual hygiene products are not necessary at all.

Natracare

The re-scripting of tampons takes on political and environmental meanings when an inventor consciously designs an ecofeminist tampon. Natracare has developed into a multinational corporation that sells organic menstrual hygiene products, including multiple sizes of tampons and sanitary napkins, as well as breast pads and incontinence pads. In 1989, a British housewife named Susie Hewson, at home with her toddler and small baby, watched a show on television called *World in Action.* Already an environmentalist and well-traveled throughout Europe, Hewson had given up commercial television and relied on alternative sources for her news information. This particular show, however, spurred her into action. The show documented how dioxins caused more than just invisible pollution in the sky and water, but affected creatures, causing hermaphrodism in fish and reptiles, with clear carcinogenic effects in humans. Importantly, dioxin can be found in bleached tampons that can then be absorbed through the delicate lining of the vagina. The more she thought about it, the more infuriated she became about the grip that corporations had on her choices and their lack of concern about her health. As she put it, "I was angered by the view that it's a marginal problem. They really didn't care. They control the market and supply chain through margins and promotions and advertisements," which results, she says, in seeing consumers as "people without power" (interview by author, September 6, 2007). And, in terms of marketing and pricing, there is "a great downward pressure on women to conform to what the manufacturers and brand owners want us to buy." Unable to sleep because of the outrage and frustration she felt toward these corporations, she found herself in the bathroom at 1:00 A.M. when she had an epiphany. "I told my husband, 'I'm going to make my own product.'" As a graphic designer, she said she knew how to solve a problem and do research because she had a good education. She believed that she had the power to "liberate herself" and as an environmentalist, she would not "turn my back on problems" (interview by author, September 6, 2007).

Living in Sweden for a number of years introduced Hewson to new ideas and

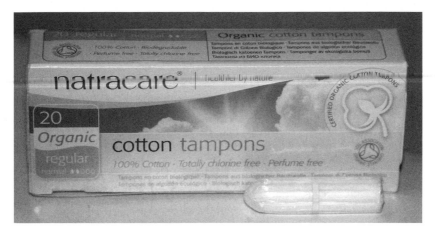

Natracare tampons. Photo by Sharra Vostral.

materials, and working in the field of design, she knew several manufacturers. She then sought out small companies with "good engineers who were willing to adapt their machinery and equipment and were prepared to work to my specifications and the raw materials I specified." In some cases, she said she had to wait a few years for materials to become available, especially organic and biodegradable components. The limitations of tampon machinery also determined the specifications of the tampons. But once the volume of sales increased, she invested in reengineering the machines to cut the organic cotton to her specifications. She also started with the view of "what I didn't want." Another core value was that the tampons had to be ecologically sound, made from renewable sources and biodegradable materials, with a low impact on the consumer in terms of allergic reactions or toxicity. In addition, she followed what she calls "the BMW principle" of motor vehicles. In the mid-1990s when some environmentally sound products were coming on the market, they were equated with low quality or less aesthetic appeal. She wanted to deliver "a high-quality product that happens to be an environmental product." She tapped into her environmental network and conducted a small trial with the organic tampons, accompanied by a questionnaire for market research, about what each user liked, disliked, and if an allergic reaction occurred. She also added questions to find out women's attitudes about products being buried at the landfill, and what kind of impact they wanted to leave based on the waste products generated as a result of their own periods. Many women, especially those suffering from contact dermatitis from tampons and sanitary pads, were thrilled to have an alternative. As Hewson put it, "[T]hey were sitting on plastic all day long," which caused vaginal irritation and rashes, and it was no wonder her organic product provided them some relief.

Her path to business success was not without obstacles, however. In 1991, as she sought out suppliers and manufacturers, she recalled there was "a lot of push and pull" to get the design she wanted premised on the materials that she specified, because many engineers were unable to envision the eco-friendly tampon. In addition, she felt that "many [of the suppliers] worked against me" and that things were done "counter to my interest to undermine the product in the market." She later learned that one of her manufacturers was "under duress from another manufacturer" to force Natracare brand from the market (interview by author, September 6, 2007). She also received threats from another arena. Her company was an early adopter of Web site technology, and as such, it posted articles and news releases related to dioxin, toxic shock syndrome, and women's health in this purposely public forum. Due in part to this educational leg of her company's mission to change women's health through better environmental choices, competitors could read about the company, too. As such, she recalled receiving "constant legal harassments that I had through letters to try to stop me from talking about issues that were relevant to the industry, whether it was toxic shock syndrome or it was chlorine bleaching." She felt that it was "par for the course—that harassment was part of the armory of corporations to silence people who want to talk about the way things really are."

Nonetheless, Hewson remains remarkably upbeat and positive about the importance of producing a product based on ethical decisions about women's health, the environment, and business practices. The company only uses products grown and derived in Europe, so transportation costs for raw materials are low because the products are manufactured there as well. She says, "[W]e know our employees, who are mostly women" and they earn an "advanced minimum wage" or a living wage. But, she also believes that so many women are personally invested in the product and tenets of the company that "it belongs to women now" and is no longer really her own. However, she has had a large hand in setting the tone for women to feel that kind of ownership. Hewson believes that "people have the ability and capacity to think about issues if they're taught and given the opportunity and given information which they can work with." As an environmentalist she learned that there were many ways to affect the system—through protest, petitions, or zoning—but the real key was to have an alternative, so people could "make a different choice" instead of doing the same old thing. By recognizing women's "humanity as consumers," she believes that "women have become aware and likely to react in more disruptive ways," which can lead them to "think about and make changes about what they buy and use." Hewson's vision to re-script a tampon with ecofeminist intent has had international impact (with its most recent product rollout in Korea in August of 2007), and the company serves as a model merging corporate, feminist, and environmental issues into product design.

DESIGN INNOVATION: VIPON

An even more radical innovation is found in Vipon, a vibrating tampon that has re-scripted the meaning of the tampon to be a technology for quelling menstrual cramps, with a camouflaged additional use as a vibrator. Steve Kilgore, a regional manager for a communications company, got the idea to put a small motor— the same as for a cell phone vibration setting—into a tampon on the day his wife exhibited what he perceived as classic signs of menstruation: short temper, agitation, and anger. When his pager vibrated, he claims his wife yelled at him about being preoccupied with his job. As he reached for the pager, he noticed a box of tampons lurking in the grocery bag resting on the table. He wondered if a vibrator, like the one inside his pager, placed in a tampon might relieve his wife's cramps ("A Vibrating Tampon" 2002). He hollowed out a tampon and inserted a tiny motor purchased for one dollar at a nearby electronics store. After a few tries, he had a prototype that hummed softly. His wife tested the new tampon during her next period, and she was amazed that it relieved her cramps. She named it Vipon, short for vibrating tampon. Enthused, Kilgore cobbled many more together, which his wife distributed to coworkers at the insurance company where she worked. The users reported their satisfaction and urged him to sell the special tampons widely.

Kilgore took this charge seriously. He applied and received two patents for Vipon (U.S. Patent 5,782,779, 1998; 6,183,428 B1, 2001). According to his patent, "[T]he invention comprises a tampon with a built-in vibrating mechanism which may assist a woman by easing menstrual cramps" (6,183,428 B1, 2001). His first patent design placed the battery source inside the tampon, with the mechanism "housed inside of a non-toxic polyethylene plastic tube which is ultrasonically welded together providing a liquid-proof container," thus preventing menstrual blood from short-circuiting the device. The tampon would be used in a similar manner to other tampons, but "the user would activate the unit by pulling the pull-string," which would start the battery. Like Tampax, Vipon is delivered in a cardboard applicator, but the battery unit in the Vipon prototype only lasted 20 minutes (5,782,779, 1998). This shortcoming, as well as the more dangerous safety concern of electrical shock to the vagina, forced Kilgore to redesign and repatent Vipon. In the second rendition, the power source is external to the device. This, however, may cause a different problem because it would be uncomfortable to have a battery dangling between the legs. In the second design, "the equivalent of the tampon removal string comprises an insulated pull cable element that encapsulates a pair of electrical leads which operatively connect the remote power source to the internal vibrating motor which is surrounded by the tampon" (6,183,428 B1, 2001). The new low-voltage battery produces power for upward of one hour.

Kilgore remains committed to the feasibility of this technology. In December 2006, his company, Another Way Products, set up a clinical trial to test the efficacy of the vibrating tampon compared to ibuprofen to quell cramps. Sponsored and conducted at the University of Kansas Department of Obstetrics and Gynecology, the trial, entitled "A Comparison Study of Pain Relief from Dysmenorrhea between the Vipon Tampon and Ibuprofen," examines complaints from fifty-one women about cramps, back pain, and abdominal pain during a four-month cycle

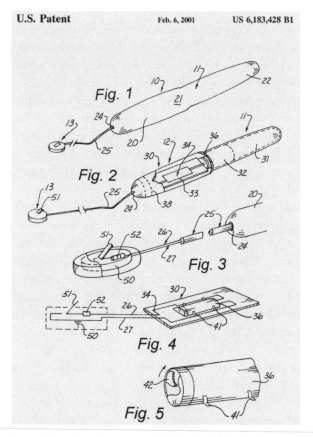

Patent for a vibrating tampon, with external battery. Steve Kilgore, United States Patent Office, 6,183,428 B1, February 6, 2001.

of a woman's period compared to a drug regime of ibuprofen to dull menstrual pain (Calkins 2007). Because the primary goal was to assess the pain relief provided by Vipon, "the subject measures her pain on the modified Melzack-McGill numerical pain scale questionnaire at various times" (Calkins 2006). Developed by psychologist Ronald Melzack, who studied "phantom limb" syndrome, the test categorizes pain into severity by ranking its intensity, as well as identifying its type—for example, searing, throbbing, etc. In addition, the "MOS-SF36 Quality of Life Questionnaire will be used pre-dose and eight hours after treatment." The Medical Outcomes Study—Short Form 36 (MOS-SF36) includes thirty-six questions grouped into eight categories such as vitality, bodily pain, and social functioning to measure the patient's overall feelings of general health. By assessing the menstruants' perceptions of pain and how quickly the pain dissipates using the Vipon or ibuprofen, the trial sought to establish scientific data statistically validating improved health. Without the data, Kilgore must rely on the qualitative testimonials provided by friends and family members.

The selected testimonials that appeared on the product Web site were lavish in their praise for Kilgore and his tampon. Not only did it eliminate cramps during the menstrual period, but for some women with endometriosis, it seemed to eliminate uterine pain as well. Penny from Lone Jack, Missouri, wrote, "[I]f you were to ask me what I personally think of the Vipon Tampon, I would tell you it's the greatest invention for Women today." She called it "a product which gives pain relief without drug side effects, for all types of feminine discomforts" ("Testimonials"). This invention designed for women was re-scripted in its purpose and design, and empowers women to practice pain management without the side effects of drugs.

However, potential users were not sold on the somewhat naive representation put forth that the tampon served a medical purpose of palliative care. The market representative for Another Way Products, Stacy Barbiere, has fielded plenty of questions about its function as a vibrator, a technology of orgasm. She says the Vipon pulsates and does not really vibrate, that "it's an active tampon for an active woman," and has absolutely no sexual benefits ("A Vibrating Tampon" 2002). Astute bloggers at plastic.com played with the potential consequences of a tampon able to induce orgasm, which, by the way, was a charge leveled against Tampax back in the 1940s (Dickinson 1949). One person noted, "[S]o instead of being able to tell which women at the office are having their periods because they are irritable, anxious, and emotional, you'll now be able to tell because they have constant dreamy satisfied smiles on their faces? Sign me up" ("Meet Vipon"). Another joked, "I can see it now. . . . Hundreds of working women across the country staring off into space with big dopey grins. Nothing gets done for 5 days of every month. The lines to the ladies [*sic*] bathroom triples in size as women rush to refresh their 'pulsating' cramp relief." The bloggers expressed great skepticism about the scripted and desexed use of the tampon, and were dubious of the testimonials, which they read as inauthentic. They instead scripted a very sexualized and orgasmic reading of the technology. The bloggers also argued that exercise, massage, heat, and sex were more viable options to manage cramps, and probably preferable to an electronic device being worn inside the vagina. Yet the use of the Vipon technology has feminist potential in terms of reducing pain or providing sexual pleasure.

NOT "FEMININE PROTECTION" BUT FEMINIST PROTECTION

Other feminists have embraced and re-scripted the tampon as a symbol of liberated womanhood; they are women who are not embarrassed by their periods or the technologies they use to manage them. In one case, the band Yeastie Girls (a pun on Beastie Boys) performed Beethoven on a panpipe flute made from tampon applicators (Yeastie Girls 1989). This was an unexpected instrumental improvisation during a concert, and heard on their 1989 album *Ovary Action*. Other examples illustrate how some women strategically use the patriarchal disgust with menstruation to their own ends. Virginia Eubanks, a colleague in the Department of Women's Studies at the University at Albany developed an interactive Web site called Brillo while in graduate school during the late 1990s. She offered a monthly tampon tip, such as "leave a tampon on the dashboard of your car to deter auto theft" (Eubanks). As the cultural logic goes, a tampon would be so disgusting to a presumably male thief that it would not be worth the effort to steal or break into the car. Another suggested that when going through a security check with a purse or backpack, leave a tampon on top to discourage further rummaging through the bag. In these instances, women's female-specific technologies are reappropriated, re-scripted, and turned on their head. They are not about absorbing shameful menstrual blood, but empowering women to fight crime or protect or convenience themselves by exploiting male fears and revulsion toward women's "feminine protection" or "feminine hygiene products."

RAPEX: THE RAPE PROTECTION TAMPON

An entirely different tampon, Rapex, re-scripts tampon as protective technologies in an entirely new context. Appalled by the contemporary rape culture in South Africa, Sonette Ehlers began thinking about how women could deter rape instead of waiting for governmental intervention. In 1998, the South African Law

Commission estimated 1.69 million rapes among a population of about 48 million, with virgins and small children victimized just as equally as adult women. Working in the field of blood matching for blood transfusion patients, Ehlers heard many tragic stories about rape. Over the course of her career, vivid comments from clients about wanting "teeth down there," presumably to deter rapists, conjured images of the fearsome vagina dentata. She also overheard stories of men's penises getting caught in pants' zippers. She put the two ideas together to design the anti-rape tampon, which she named "Rapex" (Evans 2005). Housed in a soft plastic-shaped tampon, Ehlers designed this defensive instrument as a rape deterrent worn inside the vagina, with burr-like barbs hooking into the perpetrator's penis when pressure-released. The burrs embed into the penis, with surgery the only means of removal. Because the barbs require medical extraction, Ehlers wants legislation requiring physicians to report the incident, much like that of a gunshot wound, in order to bring charges against rapists. In effect, hers is a technological fix linked to policy making to quell violent rapes. Rapex has it critics, who cite it as a torture device as well as concession to the endemic rape culture (Clayton 2005). Ehlers claims that the tampon has the potential to deter rape and punish rapists.

Rapex is an example of how women can be technologically empowered through an artifact. The Rapex tampon was consciously designed to benefit women, albeit at the expense of a man's mutilated genitalia. Although some might argue that the device is about revenge, thus that it cannot really be considered a feminist design, it certainly does benefit any woman at the moment of a rape and easily falls into the category of feminist design as outlined by Deborah Johnson in her introductory essay (this volume). Furthermore, because the technology is linked to policy, the device has the potential to challenge the culture in which rape is an assumed male privilege. There are certain potential downsides to using Rapex—rapists may probe the vaginal canal to set off the springs, product defects may injure women, and the daily wear of the tampon assumes the certainty of a rape attack—but the very flurry of interest in and horror of the device suggest that cultural disruption of rape could occur by the use of Rapex. If nothing else, it is a provocative technology to push the conversation and policy toward criminalizing rapists.

∽

In their original commercial debut with Tampax, tampons were not designed as a feminist technology, although they were presented as liberatory. Women may have appropriated them for feminist ends, but the design itself was not scripted as feminist. Corporations have sometimes appropriated a form of palatable commercial feminism to script the technologies in advertisements for the products in the hopes of attracting a wider market. However, better designs came from users who consciously created feminist innovations, re-scripting the meaning and changing the rules of the game. In both function and construction, some tampons have been redesigned and re-scripted to better women's lives. In terms of environmental sustainability, ecofeminist design has the potential to re-script an ethical engagement between producers and users. The charge of feminism is not to just "confront technology" (Wajcman 1991) but to engage and even embrace it with critical and material discourse. Feminist designers need to cater to the multiple and varying desires of women in all their shapes and forms. Both men and women must adhere to feminist design tenets in every step of product development, design, product testing, and manufacturing, while women, no doubt, will continue appropriating and re-scripting menstrual hygiene technologies as they see fit.

ACKNOWLEDGMENTS

I would like to thank Eric Roth for introducing me to the Yeastie Girls.

I use the term "Rapex" to refer to the original concept of the defensive anti-rape tampon. Sonette Ehlers abandoned this design by 2006 and instead focused on an anti-rape condom and changed the spelling to RapeX. Shortly thereafter, the condom became known as Rape-aXe, not to be confused with RAPEX, the EU warning system against dangerous goods in the marketplace. Thus, for historical accuracy, I utilize the original Rapex spelling and refer only to her first tampon design.

References

Akrich, Madeleine. 1992. "The De-Scription of Technical Objects." In Wiebe E. Bijker and John Law, eds., *Shaping Technology/Building Society: Studies in Sociotechnical Change*, 205–24. Cambridge, Mass.: MIT Press.

Bailey, Richard. 1987. *Small Wonder: How Tambrands Began, Prospered and Grew.* Palmer: Tambrands.

Brodie, Janet. 1994. *Contraception and Abortion in Nineteenth-Century America.* Ithaca, N.Y.: Cornell University Press.

Calkins, John. 2007. "A Comparison Study of Pain Relief from Dysmenorrhea between the Vipon Tampon and Ibuprofen." Electronic document. http://clinicaltrial.gov/show/NCT00456079. (Accessed September 4, 2007.)

Clayton, Johnathan. 2005. "Anti-Rape Device Must Be Banned, Say Women." *Times* (June 8). Electronic document http://www.timesonline.co.uk/tol/news/world/article531013.ece. (Accessed April 14, 2007.)

Dickinson, Robert Latou. 1949. "Sanitary Pads and Tampons." *Consumer Reports* (August): 352–55.

"Early Ad for o.b. Tampons." N.d. Advertisement. Electronic document. http://www.mum
.org/obger50s.htm. (Accessed August 14, 2005.)

Eubanks, Virginia. N.d. "Tampon Tip." *Brillo Magazine*. http://www.brillomag.net/No2/
contents.htm. (Accessed May 22, 2006.)

Evans, Jenni. 2005. "Inventor of Rape Device Prepares for Launch." *IOL News for South
Africa and the World* (August 30). Electronic document. http://www.iol.co.za/index
.php?set_id=1&click_id=15&art_id=qw1125413643318B261. (Accessed April 14, 2007.)

Farrell-Beck, Jane, and Laura Klosterman Kidd. 1996. "The Roles of Health Professionals
in the Development and Dissemination of Women's Sanitary Products, 1880–1940."
Journal of the History of Medicine and Allied Sciences 51 (July): 325–52.

"Geschichte des Tampons." N.d. *Frau TV*. Electronic document. http://www.wdr.de/tv/
frautv/archiv2002/f240402_3.html. (Accessed June 26, 2007.)

Gilbreth, Lillian. 1927. "Report of Gilbreth, Inc." Papers of Frank and Lillian Gilbreth,
Special Collections, Purdue University, N-File, Box 95, 20–21, p. 70.

Michael L. Kehm v. Proctor & Gamble, United States Courthouse, Cedar Rapids, Iowa,
April 5, 1982.

McGaw, Judy. 2003. "Why Feminine Technologies Matter." In Nina E. Lerman, Ruth Oldenziel, Arwen P. Mohun. eds., *Gender and Technology: A Reader,* 13–36. Baltimore:
Johns Hopkins University Press.

"Meet Vipon—The Cramp Quelling Tampon." N.d. *Plastic*. Electronic document. http://
www.plastic.com/article.html;sid=02/08/24/15042627. (Accessed September 4, 2007.)

Oudshoorn, Nelly, and Trevor Pinch, eds. 2003. *How Users Matter: The Co-Construction
of Users and Technology.* Cambridge, Mass.: MIT Press.

"Sanitary Protection without Pads! Pins! Belts!" 1935. Advertisement. *New York Times*
(September 8): SM18.

Testimonials. N.d. Vipon. Electronic document. http://www.vipon.com. (Accessed December 6, 2002.)

"A Vibrating Tampon to Relieve Menstrual Cramps." 2002. *Pitch*. Electronic document.
http://www.pitch.com/issues/2002–08–22/stline.html/1/index.html. (Accessed September 4, 2007.)

Vostral, Sharra L. 2008. *Under Wraps: A History of Menstrual Hygiene Technology.* Lanham,
Md.: Rowman and Littlefield/Lexington.

Wajcman, Judy. 1991. *Feminism Confronts Technology.* College Station: Pennsylvania State
University Press.

"Welcome This New Day for Womanhood." 1936. *The American Weekly* (July 26). Electronic
document. http://www.mum.org/tamad36.htm. (Accessed November 4, 2007.)

Yeastie Girls. 1989. *Ovary Action*. http://www.myspace.com/yeastiegirlzofficial. (Accessed
February 24, 2009.)

6 From Subaltern Alignment to Constructive Mediation

Modes of Feminist Engagement in the Design of Reproductive Technologies

ANITA HARDON

THIS CHAPTER DESCRIBES how women's health advocates contested the anti-feminist-nature of new reproductive technologies, highlighting subsequent changes in its use and design. At stake in this chapter are the gender scripts (Oudshoorn et al. 2002). The concept of gender script indicates two processes at work in the mutual shaping of gender and technologies. First, technologies become gendered because designers anticipate the gendered interests, skills, motives, and behavior of future users. These gendered representations of future users are as it were "inscribed" or in other words materialized into the design of the new product (Akrich 1992; Oudshoorn et al. 2002; van Kammen 1999). Second, these gender-inscribed artifacts shape and define the agency and power of women and men.

I first describe how and why women's health activists criticized the gender scripts of two new longer-acting contraceptive technologies, Norplant and the antifertility vaccines. They claimed that the artifacts, designed primarily with population control aims in mind, disempower women (see also chapters 2 and 4, this volume, which outline how menstrual-suppressing birth control pills and breast pumps have been criticized for their potential antifeminist effect).

I not only describe the contestations, but also point out how the controversies led to changes in the gender scripts of these two technologies and to greater involvement of women in the design of future reproductive technologies. The latter is illustrated by a description of constructive collaboration between women's health advocates and researcher scientists in the development of microbicides. Microbicides were designed to protect women from HIV infection.

I also reflect on the role I played in these controversies as a participant observer. I, the author, witnessed how the campaigns to ban Norplant and to stop the research on antifertility vaccines took place and participated in consultations on their use and acceptability. I was subsequently asked to develop methods to explore acceptability and desired attributes of microbicides. In each of the cases, my role was different. I outline three different modes of engagement:

subaltern alignment, reflexive dialogue, and constructive collaboration (see the sections below).

To write this chapter, I retrospectively reexamined the campaign materials and other texts from women's health advocacy groups, including the correspondence with research institutions. Until 1992, I worked part time as the research coordinator for the Amsterdam-based Women's Health Action Foundation (WHAF). For the majority of my time, I held the position of medical anthropologist at the University of Amsterdam. Throughout the controversies I wore both hats: participating in the (re)design processes as an activist and taking meticulous notes on the actions and reactions as a researcher. I have remained involved in health-related activism, always balancing the roles of activist and the reflexivity involved in research. In addition to describing the gender-(re)scripting processes, I reflect on my own roles in each of the case studies, and in doing so, on my own roles in feminist design.

SETTING THE SCENE

The women's health advocacy movement is a broad movement with a variety of roots. The movement includes radical feminist groups calling for abortion rights, women's health organizations opposing medicalization of women's reproductive functions, community-based health groups, and feminist researchers and journalists. Women's health advocates would generally agree that they share a goal of empowering women to control their own fertility and sexuality with maximum choice and minimum health problems. They also have a common skepticism toward medical claims about the safety of the reproductive technologies.

In the industrialized world, the movement finds its origins in "second-wave" feminist movements, which in the late '60s and early '70s rallied around the right to contraception and abortion and to express a growing concern about patriarchal control in medicine. Free contraception and abortion on demand were seen to be keystones of women's liberation (Berer 1997). In 1980, the International Contraception, Abortion, and Sterilization Campaign (ICASC) "Women Decide!" was launched. The members of ICASC in 1980 included feminist groups and networks in Europe, Latin America and the Caribbean, Africa, North America, India, Australia, and New Zealand (ICASC 1980). The ICASC later changed its name to the Women's Global Network on Reproductive Rights (WGNRR) (Keck and Sikkink 1998).

ICASC organized the fourth International Women and Health meeting in Amsterdam in 1984; this changed the tone of international women's health activism. This meeting had as its slogan "Population Control—No Women Decide!" The participants at this meeting were concerned about Norplant, the new long-acting contraceptive technology that appeared to be designed more for the purpose of population control than for the purpose of women's health and reproductive rights (Hardon 2006). The meeting challenged the rationale on which population programs aimed at reducing fertility in developing countries were based; namely, that limiting family size as a societal responsibility has precedence over individual well-being and individual rights. There were reports on the way state population programs in countries such as India and Bangladesh aimed at reducing population growth by means of coercive family planning programs. Mistrusting their governments, women expressed concern about the safety of the new longer-acting contraceptive technologies that were being made available in target-oriented family planning programs (Garcia-Moreno and Claro 1994; Hartmann 1995). By framing their concerns in relation to reproductive rights, the women's health advocates had found an oppositional collective identity and a powerful counterdiscourse to that of "population control," as defined by the World Population Plan of Action (WPPA) and adopted at the 1974 Conference on Population and Development in Bucharest, and the subsequent 1984 International Conference on Population held in Mexico City (Hardon 2006).

In the second half of the 1980s, the global women's health movement gained momentum. Issues of concern relating to contraception were the adverse effects of the Dalkon Shield (which led to litigation against its manufacturer and a successful claim for compensation), the distribution of the hormonal injectable Depo Provera in family planning programs in the South (at the time the technology was still not approved for distribution in the United States), and the safety of the new long-acting hormonal implant Norplant (Hardon 1992). By the mid-1990s, the women's health movement had become a strong global movement with several transnational advocacy networks.

The controversies around the safety of new reproductive technologies, initiated by the women's health advocates, resulted in an increased commitment shown by international research institutions such as the Human Reproduction Program of the World Health Organization (WHO) and the U.S.-based Population Council to involve women's health advocates and potential users in the setting of research priorities and in decision-making mechanisms concerning the design of reproductive technologies. It is important to realize that these design agencies are public. We have seen in chapter 4, the breast pump case, how the private-sector manufacturers of these technologies were all along interested in the needs and interests of women. A wide range of products catering to different kinds of users is on the market. Private-sector manufacturers know that they need to provide choice to expand the markets for their products. WHO and the

Population Council initially didn't see women as future users of their products, but instead publicly funded family planning programs, which are happy with a one-size-fits-all technology because it simplifies delivery.

Initiatives illustrative of the trend toward greater involvement by women's health advocates in the contraceptive development process include the 1991 meeting organized by WHO with the International Women's Health Coalition (IWHC) on "The Selection and Introduction of Fertility Regulation Technologies" and the declaration from a symposium on "Contraceptive Research and Development for the Year 2000 and Beyond" attended by reproductive researchers, managers of programs, and women's health advocates. Among other things the symposium recommended:

> Women's health advocates and potential users should be represented in all decision-making mechanisms and advisory bodies that are established to guide the research process, including definition of criteria for safety, determination of research priorities, design and implementation of research protocols, setting and monitoring of ethical standards, and decisions on whether to pursue a fertility regulation method from one stage to the next, especially decisions to move from clinical trials to introductory trials, and from introductory trials to the introduction of a method into family planning programs. (Anonymous 1993)

SUBALTERN ALIGNMENT: THE NORPLANT CASE

What kind of technology is Norplant, and why were women's health advocates so concerned about its introduction in the 1980s? In what ways did the women's health movement engage in the "re-scribing" of this new reproductive technology? Norplant is a hormonal contraceptive consisting of five levonorgestrel-releasing rods. After surgical insertion in a woman's upper arm, it works for a period of five years. Women's health groups in Bangladesh, Thailand, and Brazil became aware of this new technology when the New York–based Population Council (with funding from the U.S. government and private foundations) began a global program of multicenter clinical and acceptability trials involving forty-four developing and developed countries in the early 1980s (Correa 1994). At the national level, the women's health groups were initially concerned with unethical trial conduct. At the 1984 International Women and Health meeting (see above), women's health advocates working for a local nongovernmental health collective called UBINIG in Bangladesh reported how women were recruited to participate in the trials by means of advertisements published in *Holiday Weekly* on October 4, 1981.

The advertisement does not inform readers that women will participate in an introductory trial and describes this experimental technology as a "wonderful innovation." UBINIG found that family planning workers recruiting women for the trial in Bangladesh told women that Norplant would bring women happiness. They discredited other methods (Akhter 1995). Mobilized by UBINIG, a petition was sent to the Ministry of Health signed by 150 concerned health workers. The Bangladesh trial was subsequently postponed. In Brazil, also, as a result of irregularities in trial conduct and protests from women's health advocates, the Norplant trial was stopped. There was specific concern about failures in informed consent procedures (Correa 1994).

In 1989, women's health activists from Bangladesh, Brazil, India, Indonesia, Thailand, Denmark, Finland, and the Netherlands met to discuss design issues related to Norplant and other new contraceptive technologies at a seminar sponsored by the Amsterdam-based Women and Pharmaceuticals Project, for which I worked as a part-time researcher. In response to the concerns of women's health advocates that Norplant was designed to control populations rather than to protect women's health and rights, I had reviewed the clinical evidence on Norplant. My findings confirmed that the designers had neglected possible negative health effects of this new reproductive technology. I was specifically concerned about the way the researchers labeled menstrual disturbances as minor side effects and the lack of recognition of potential safety problems when used long term and during pregnancy (either as a result of method failure or because of insertion without checking for pregnancy).

Participants at the meeting decided to do field studies to learn more about the way women experienced the menstrual disturbances caused by the methods, and more generally to observe the way in which Norplant was being introduced in different countries. I initiated and supported the field studies, which were conducted from 1989 to 1991, coordinated by the Women and Pharmaceuticals Project. The findings of the studies were published in the book *Norplant: Under Her Skin* (Mintzes et al. 1993). The collaborating action-researchers had found that the menstrual disturbances were perceived to be a problem in many of the research settings, given that delay or absence of menstruation is considered unhealthy in many societies. We found that health workers gave women who suffered from menstrual bleeding in Finland and Indonesia estrogens or Vitamin K. We considered these inappropriate, arguing that these were ad hoc treatments using pharmaceuticals for purposes for which they had never been tested and that they constituted a further medicalization of healthy women following Norplant use. The field studies also showed that in Indonesia, Thailand, and Brazil, women had difficulty in having Norplant removed. In some cases, removal was refused by health workers. In other cases, the removal process itself was difficult because of incorrect insertion and/or the breaking of rods during removal. The lack of access to removal services indicated in our view the disempowering nature of the technology. In a review of acceptability studies, we concluded that the scientists had failed to consider women's own views on the acceptability of side effects.

Based on these field studies, we defined conditions that would need to be met for the method to be used. These included:

1. The recognition of women's reproductive rights in the provision of Norplant, specifically the need to provide women with a choice of methods and informed consent when considering Norplant insertion;
2. The need to avoid medicalization of side effects;
3. The need to consider the effects of menstrual disturbances on women's day-to-day lives;
4. The need for sterile administration;
5. The need for access to safe removal services;
6. Adequate follow-up care, and;
7. Appropriate provision of information in the media.

The Women and Pharmaceuticals Project engaged in a dialogue with WHO and the Population Council on the need for good quality of care in providing the methods. We chose to confront the potentially disempowering gender script of the technology by formulating guidelines for its actual use. Our attention shifted from the hardware—the contraceptive itself—to the software—the arrangements through which the technology was to be made available, which elsewhere I have called the introductory script (Hardon 2006). Good quality care, in our view, could enable users to select the particular contraceptive technology that best fit their day-to-day lives.

The book *Norplant: Under Her Skin* was reviewed positively in the press but criticized heavily by the designers of Norplant, the Population Council. Karen Beattie, responsible for the Norplant introduction program, wrote that the analysis was skewed and that its conclusions on the need to avoid medicalization were "truly amazing" (1993 letter, K. Beattie to R. Petchesky). In an attempt to attack our credibility, we were said to lack understanding of clinical research and practice. An intensive correspondence followed in which we responded to each of her criticisms (1993 response, A. Hardon to K. Beattie).

While WHAF activists were negotiating with WHO and the Population Council on the conditions for safe use of and the safety claims for Norplant, other women's health activists were developing a different strategy. The problems with Norplant were discussed at the International Conference on People's Perspectives on Population, held in Bangladesh in December 1993 (a meeting organized by

UBINIG and the Third World Network) and attended by women from thirty-four countries. The symposium concluded in the *Declaration of People's Perspectives on Population*:

> Increasingly technologies are invented that are controlled by the providers, that is, the physicians, the drug companies, the state. Formerly, contraceptives, like the diaphragm, were more under the control of women. Whether in relation to curbing or enhancing fertility, these provider-controlled technologies effectively undermine women's control over their lives while burdening them with full responsibility for fertility and absolving men of their responsibility. Therefore, long-acting contraceptives such as Norplant are not an advance in contraceptive technology, but an advance in population control. They are purposeful instruments inspired by eugenicists whose programs of population control were designed explicitly to curtail the number of black, indigenous, disabled and poor white peoples. (Akhter 1995, 105)

This anti-Norplant alliance lobbied for their views in the process leading up to the 1994 International Conference on Population and Development (ICPD). UBINIG and the (radical) Feminist Network for International Resistance to the New Reproductive Technologies and Genetic Engineering (FINNRAGE), along with other groups, distributed the leaflet *No to Norplant* to the delegates of the preparatory conference in New York.

It stated:

> Norplant is a provider-controlled and an inherently coercive method. . . . We urge governments, the UN bodies and women's groups all over the world to call for the withdrawal of Norplant from the family planning programs from all the countries and to provide safe and user-controlled methods of fertility regulation in concert with broad-based health care programs. (Akhter 1995, 108)

By 2001, Norplant had been registered as a safe and effective contraceptive method in sixty countries, but was not used widely. It is estimated that only around 10 million women worldwide use implants, accounting for around 4 percent of contraceptive users (Bongaarts and Johansson 2000).

Low rates of Norplant use are probably the main reason that manufacturers have withdrawn Norplant in two countries: the United Kingdom and the United States. In both countries it had sparked mass legal action from women who claimed to suffer side effects, ranging from nonstop bleeding to hair loss to suicidal depression.

Reflecting on my involvement in this controversy, the mode of engagement in the controversy was that of *subaltern alignment*: as researcher I aimed at putting forward the problems the technology posed for women users in situations of relative powerlessness (i.e., those faced with coercive family planning programs). The aim of the Women and Pharmaceuticals Project, for which I worked, was to confront the authoritative medical-demographic arena involved in the design of the technology with the lived experiences of women, whose access to health care is constrained by power differentials between health workers and patients. Bias and blind spots in mainstream science were critiqued. This critique was multidisciplinary. It dealt both with the biomedical efficacies of the technologies and with their social uses and abuses. Our critique stemmed from listening to women living in a variety of contexts all over the world. Unlike some of the more radical women's health groups (such as FINNRAGE), I was not convinced that the technology was inherently oppressive—that is, antifeminist—and our informants who had experience in using the technology didn't feel so either. The women's health organizations in which I participated called for a change in the introductory scripts of the technologies; that is, in the systems through which they were to be made available to women. We argued that such systems should ensure informed consent prior to insertion, sterile administration, and adequate follow-up, including removal on demand. We felt that the technology could meet women's needs—that is, empower women—if these conditions were met.

REFLEXIVE DIALOGUE:
THE ANTIFERTILITY VACCINE CASE

With Norplant, women's health advocates raised concerns about the introduction of the method in family planning programs, its disempowering potential, and misleading claims on safety and acceptability (of menstrual disturbances) to users. When the women's health advocates became aware of the technology, it was already on the market in many countries. Field studies allowed women's health advocates to present researchers with actual problems with the new reproductive technology, experienced by women in everyday life. The contestations led to a re-scribing of the gender scripts for the hardware of this technology, as we will see. And, as was the case for Norplant, the introduction scripts (the scripts of the systems by which the artifacts are made available to women) changed.

The antifertility vaccine case history is different because, unlike Norplant, the antifertility vaccines were contested when they were still in the process of being tested in clinical trials; that is, the gender scripts were still under construction and more open to adaptation. The contestations led to the halt of trials on the prototype that was seen to be most antifeminist by the women health advocates and to the construction of new prototypes with more empowering scripts.

The idea of regulating fertility by immunological means has its origins in the science of reproductive immunology, which showed that conception and embryo implantation can be interrupted by immunological manipulation (Barzelatto 1991; Jones 1982). In the late 1970s, the International Committee for Contraception Research of the Population Council initiated human clinical pharmacological studies with an anti-hCG vaccine in Sweden, Finland, Chile, and Brazil (Nash et al. 1980). The hormone hCG (human chorionic gonadotropin) is produced in pregnancy. Its role is to prevent the disintegration of the corpus luteum of the ovary and thereby maintain the progesterone production that is critical for a pregnancy. Following these early experiments, two Phase 1 clinical trials (small-scale safety trials on volunteers) were conducted in the 1980s: anti-hCG vaccines were tested under the auspices of the Population Council in India and Scandinavia with a total of eighty-eight surgically sterilized women (Talwar et al. 1990), and a WHO-sponsored trial testing the safety of an anti-hCG vaccine was conducted in Australia with thirty surgically sterilized women (Jones et al. 1988).

In their published reports on these studies, the researchers refer to their work as part of an important solution to the global population crisis, a project that is distrusted by women's health advocates because they fear that in designing the technology, the researchers will favor longer-term efficacy over safety of the technology. Two researchers from the New Delhi National Institute of Immunology (NII), for example, wrote:

> Most conservative estimates predict human global population to cross six billion by the end of the 20th century. . . . It poses a major challenge for developing countries and demands mobilization of additional resources . . . to maintain the complex relationship between growing population and environment. To overcome this problem it is pertinent to evolve new safe and effective contraceptive agents. Vaccines for immunocontraception are an interesting proposition as it will be cost-effective, and most developing countries have infrastructure for the appropriate delivery. (Gupta and Koothan 1990)

Reacting perhaps to the controversy that was arising around safety and acceptability of Norplant, as discussed above, WHO called for careful evaluation of this new longer-acting reproductive technology. The WHO biennial report on human reproduction stated that antifertility vaccines should:

> [h]ave long-lasting protective effect after a single course of immunization; they would not cause menstrual-cycle disturbances and other hormone-dependent side-effects; they would be easy to administer by a well-accepted procedure; and they could be manufactured at low unit cost. (WHO 1990)

With Norplant, the designers formed a "united front" in their assessment that the technology was safe to users. With the antifertility vaccines, the designers differed in their assessments of acceptable risks. Two distinct prototypes of anti-hCG vaccine were emerging, one that I have characterized elsewhere as maximizing efficacy, the other as maximizing safety (Hardon 1997). WHO's HRP chose to maximize safety by selecting the small section (peptide) of the beta subunit of hCG as an antigen with possible drawbacks in terms of efficacy. Talwar and his colleagues at the NII in India, on the other hand, did not select an antigen that is unique for hCG; instead, they used the whole beta-hCG subunit, which is in some ways similar to the LH hormone that regulates the menstrual cycle. The vaccine theoretically could also disrupt LH (i.e., cross-react) and thus cause women to have menstrual disturbances. By choosing the whole beta-hCG subunit as their antigen, they disregarded this risk. The 1978 guidelines of the WHO Task Force on immunological methods advised against selecting antigens that carry a risk of "cross-reactivity" (WHO 1978). The Indian researchers argued for "a moderate degree" of cross-reaction with LH, as long as women menstruate normally, in order to enhance the effectiveness of the vaccine (Talwar et al. 1988).

The chair of the WHO Steering Committee of the Task Force on Vaccines for Fertility Regulation—of which Talwar was also a member—commented on this issue in a review article:

> With regard to the relative merits of the two types of vaccine described above, it is still too early to reach a conclusion. Quite possibly, both vaccines will find their place in the armamentarium of contraceptive agents. The cross-reactions elicited by the intact beta chain vaccine are worrying, but that concern diminishes as the number of women who have been vaccinated without adverse consequences increases. (Mitchison 1990, 726)

Initial questions from a women's health perspective about the safety and acceptability of antifertility vaccines were raised at the 1989 WHO symposium on the safety and efficacy of vaccines for fertility regulation, which I attended with Judith Richter as consumer representatives on behalf of Health Action International (an international network of consumer, health, and development organizations). We were the only advocates invited to this meeting.

The aim of the symposium was to review aspects of present and past work on the development of antifertility vaccines, particularly relevant to the testing of their safety and efficacy (Ada and Griffin 1991). At this meeting, we were first confronted with the different views among the researchers on the relative safety features of the two anti-hCG vaccine prototypes that were emerging. Not willing to confront the Indian researchers in plenary, the researchers involved in the safer beta-hCG peptide vaccine urged us to raise questions on this issue.

In a report on the symposium published in the WGNRR newsletter, I summarized several concerns with both anti-hCG vaccine prototypes raised by Richter and myself (Hardon 1989). These included concerns about the intrinsic qualities of the vaccine, such as: (1) difficulty of "switching off" the immune response (the temporary irreversibility is a problem for women who experience side effects, such as menstrual disturbance or autoimmune reactions); (2) unknown consequences if a woman is pregnant when given the vaccine; and (3) risks of cross-reactivity and allergic reactions to the immuno-carrier. In addition, there were concerns about operational issues to be considered, especially in settings where health care services are inadequate. These included: (1) the need for a test to determine whether a woman has a protective level of antigens to hCG; (2) the need for additional protection during the immunological lag period (the period after the injection and before the immune response has developed to an effective level); and (3) the vaccine's potential for abuse if distributed in coercive population programs (women could be injected with the antifertility vaccine without their consent).

In the same article, I suggested that the clinical trials conducted with the whole beta subunit were inappropriate because of the potential health risks related to cross-reactivity. We had studied the basic immunology underlying these technologies to be able to engage in these technical debates on product design. Having rephrased the women's health concerns in technical language allowed us to engage in a reflexive dialogue with the researchers.

Griffin responded in the form of a letter to the editor, with assurances that the development of the vaccine would be stopped if serious adverse effects occurred or if the vaccine was found to have teratogenic effects (causing deformities in the unborn child). With respect to the trials using the whole beta subunit, he commented that the trials conducted did not reveal the occurrence of menstrual disturbances because of cross-reactivity (Griffin 1990).

The above concerns about safety and the antifertility vaccine's potential for abuse have caused many women's health advocates to vehemently oppose this technology and to question the rationale for its development. During workshops and meetings held in the early 1990s, they commented on the supply-defined contraceptive-development process, calling for a reorientation toward user needs (Richter 1992). Schrater, a feminist immunologist, wrote a review article in which she supported our concerns (see above, Hardon 1989) about "allergy, autoimmunity, irreversibility and teratology" and possible abuse and direct or indirect coercion by the state (Schrater 1992, 47).

In line with the increased commitment to involve women's health advocates in the development of new contraceptive technologies, as asserted in the 1991 meeting (WHO 1991), in August 1992 WHO organized a meeting for researchers and women's health advocates to discuss the issues at stake concerning antifertility vaccines. In a background paper for this meeting, the manager of the Task Force on Vaccines for Fertility Regulation provided information on the state of the art as a basis for discussion. Central to Griffin's paper was the assurance that technical solutions could be found for the concerns about the intrinsic characteristics of the vaccines. More animal studies and clinical trials were needed to clarify the exact mechanism of action to develop methods to reverse the effects and to assess long-term safety. He asserted that sufficient information was available to indicate that it would be feasible to develop antifertility vaccines that were free of overt pharmacological activity and the metabolic, endocrine, and physical disturbances that often accompany other methods of birth control and could confer mid- to long-term (three months to one to two years) but not permanent protection following a single administration. The user, he suggested, would be able to select the preparation offering the length of duration desired (Schrater 1992).

This notion of reproductive choice was new. Whereas prior to the actions of women's health advocates the scientists aimed to develop a longer-acting injectable contraceptive with limited potential for so-called user failure, Griffin now proposed that users should be able to choose the length of efficacy they wanted. The nonpermanent nature of the vaccine's effect, and the possibility of reversing the effect on demand, in his view, would alleviate the consequences of abuse, should it occur. In reframing the technology as one contributing to choice rather that population control, Griffin changed the gender script of the technology.

In June 1993, nineteen women's health advocates from twelve countries met in Bielefeld, Germany, hosted by the German organization BUKO Pharmakampagne (a nongovernmental organization involved in third world health and development issues), to discuss their views on the antifertility vaccine. The meeting had an open and closed session. Griffin was invited to the open session and asked to present the scientific data on the antifertility vaccines. He was questioned at length about its safety and efficacy. In the closed session of the workshop, the women's health advocates drafted the "Call for a Stop on Research" (WGNRR 1993a).

In November 1993, the WGNRR launched the global "Call for a Stop of Research on Anti-Fertility 'Vaccines'" that was sent to research institutes and funders signed by 232 organizations from eighteen countries:

We, the undersigned, call for an immediate halt to the development of immunological contraceptives because of concerns about health risks, potential for abuse, unethical research, and the assumptions underlying this direction of contraceptive research.

... Immunological contraceptives will not give women greater control over their fertility, but rather less. Immunological contraceptives have a higher abuse potential than any existing method.

... Immunological contraceptives present no advantage for women over existing contraceptives.

... They interfere with complex immunological and reproductive processes. There are many potential risks: induction of autoimmune diseases and allergies, exacerbation of infectious disease and immune disturbances, and a high risk of fetal exposure to ongoing immune reactions.

... The concept of antifertility "vaccines" was conceived in a "demographic-driven, science-led" framework. (WGNRR 1993b)

By May 1996, this call had been endorsed by 472 groups from forty-one countries. Brazil (around 120 signatories), India (ninety-five signatories), and Germany (around sixty signatories) accounted for more than half of the responses (Yanco et al. 1996). I didn't sign the call because I didn't support the call to stop all research. I knew from our earlier studies on Norplant that women in some sociocultural settings would welcome a longer-acting method that didn't have hormone-related side effects. In my view, the unsafe vaccine based on the whole beta-subunit of hCG developed by the National Immunology Institute and the Population Council in New York needed to be stopped. Here again, we see that the more radical feminist strands of the women's health movement believed that the technologies were inherently antifeminist—that is, oppressive—and thus should be banned.

In the first half of 1994, the research institutions, such as the WHO's HRP, the Population Council, and the Contraceptive Research and Development Programme (CONRAD), sent reactions to the WGNRR, the campaign's global secretariat. In these reactions, they refer to the routine procedures in the development of contraceptives. They stated that safety and efficacy were being assessed in clinical trials and that the outcomes of these trials would resolve the issues. They also emphasized that their institutions supported reproductive rights principles, implying that the research was not demographically driven, and that the methods could have benefits to users (1994 letter, H. L. Gabelnick to B. Stemerding; 1994 letter, M. Catley-Carlson to B. Stemerding). Dr. G. Benagiano, director of the WHO's HRP, for example, stated: "I agree completely with the aim of WGNRR ... the right of women to decide whether, when and how to have children. ... It is however my contention that this aim also includes the right of women to choose what method of family planning to use, including, if they wish so, an antifertility vaccine." Furthermore, "We feel that a fully developed and tested family planning method . . . will be an attractive option for those women who wish to postpone their first pregnancy, to space births at an interval that has positive health benefits for the mother and her children" (1994 letter, G. Benagiano to WGNRR). Interestingly, these responses suggest that family planning providers were no longer envisaged as the future users of the technology, but rather women, who had the freedom to choose between contraceptive options.

In an attempt to contribute to the dissemination of nonspecialist information on antifertility vaccines, and in response to the "Call for a Stop" (Ravindran and Berer 1994), the May 1994 issue of *Reproductive Health Matters* contained an article in which the researchers involved in the development of the "safer" beta-hCG peptide vaccine reviewed the current status of antifertility vaccines (Griffin et al. 1994). Concerning the problem of abuse, they pointed to the need for improved quality of care and education; that is, improved introductory scripts. In the same issue, the director of the Women of Color Reproductive Health Forum in New Orleans suggested that each woman would have to decide for herself if the contraceptive's side effects were worth the risk (Shervington 1994). R. Macklin, a professor of bioethics, argued: "Those who would restrict women's options are being paternalistic in their attempt to curtail the freedom to choose" (Macklin 1994, 112). A. F. Schrater also added her voice by distinguishing between the two types of anti-hCG vaccines in development. She considered that the long-term risks of the prototype developed by the New Delhi National Institute for Immunology (NII) and the Population Council to be unacceptable. She supported further development of the safer alternative developed by WHO (Schrater 1994). None of these respondents supported the campaign to stop the development of all anti-hCG vaccines.

In early 1994, WHO initiated Phase 2 clinical trials on its theoretically safe anti-hCG prototype in Sweden. Of the twenty-five volunteers selected to participate in the trial, the first seven to receive the vaccine all experienced unexpected side effects, which included pain at the injection site, fever, and, in two cases, sterile abscess formation. The trials were stopped by mid-1995 (WHO 1995a). Following this experience, the Task Force on Vaccines for Fertility Regulation of WHO shifted its attention to "advanced prototypes" and "optimized vaccines." The aim was to develop a totally synthetic anti-hCG vaccine containing bioengineered immunogens and a controlled-release system designed to provide immunity of a predictable and controlled duration. An orally active formulation of this "optimized" version of the anti-hCG vaccine was also being investigated

(WHO 1995a). In aiming to develop an oral vaccine, the researchers showed that they took the critique of women's health activists on provider-controlled methods seriously.

In mid-1995, the activists involved in the No-to-Anti-Fertility Vaccines Research Campaign met again, in Canada, where they also aimed to negotiate with representatives of the International Development Research Centre (IDRC). IDRC had supported the development of the antifertility vaccine at the NII in India, and held the patent on the anti-hCG vaccine developed in India. Dr. Talwar told me in 1996 (he had then retired as director of the NII) that IDRC had decided to stop funding the development of antifertility vaccines by the NII in New Delhi. He stated, "Our research has been stopped by the women dictating . . . because they were so persistent I got a low priority" (interview with Dr. Talwar, August 1996, NII). In January 1997, a letter from the president of IDRC confirmed that funding had stopped, but said this was only because NII officials were not intending to request further funding (letter, K. A. Bezanson to K. Seabrooke). *Nature Medicine* (Jayaraman 1997) reported in May 1997 that India was indeed downgrading research on contraceptive vaccines. The new director of the NII, Dr. S. K. Basu, was quoted, saying, "We cannot allow this vaccine to enter phase III trials until its long-term safety is established" (Viswanath and Kirbat 1997).

The controversies surrounding the safety and acceptability of Norplant and the antifertility vaccines led to commitments to involve end users in the design of future reproductive technologies. In 1995 and 1996, reflection on research priorities was also taking place at WHO. In a discussion paper, the HRP put forward criteria for the program's future research and development policy practice. It stated, "The views, needs and preferences for fertility regulating methods as expressed by men and women, past, current or potential users, should guide the selection of new methods for development" (WHO 1995b, 13).

Ironically, this increased concern at WHO about the needs of users led to new campaign activities. In early 1996, the women's health advocates involved in the No-to-Anti-Fertility Vaccines Research Campaign launched an international postcard action, following an informal telephone conversation with Griffin, in which he reportedly said that the HRP would consider ending research on the anti-hCG vaccine if "potential users would not want the method" (1996 letter, B. Stemerding to all people involved in the campaign). In this postcard action, women were called upon to sign postcards addressed to Griffin, stating: "I do not support the development of immunological contraceptives. Women and men alike need contraceptives that enable them to exercise greater control over their own fertility, without sacrificing their integrity, their health, or their well-being.

In addition, the potential for abuse is simply too great with immunological contraceptives, which could easily become tools for population control." In this campaign, women positioned themselves as users and claimed a right to stop research on vaccines.

This case study shows how women's health advocates, as in the Norplant controversy, were concerned about safety (possible autoimmune reactions and cross-reactions) as well as rare and long-term risks. Part of the women's health movement (i.e., the Women's Global Network on Reproductive Rights) chose to call for a ban on the new technology on these grounds. Like Norplant, antifertility vaccines were defined by these activists as inherently antifeminist. Other women's health activists called for women to make informed choices. They acknowledged potentially liberating effects of these new technologies as long as their introductory scripts would acknowledge the potential for abuse.

Reproductive researchers called upon the women's health advocates to base their risk assessments on the uncontroversial facts that would emerge from the ongoing Phase 2 controlled clinical trials. The more radical women's health advocates aligned themselves with the trial subjects that would actually be injected with the vaccine and saw no justification for this experimentation. Women's health advocates fundamentally questioned the need for a contraceptive method that meddled with the immune system and disagreed with the empiricist proposition that risks needed to be tested in trials.

The antifertility vaccine controversy led some reproductive researchers to change the gender scripts of the technologies that they were designing. They no longer aimed at empowering family planning workers with a tool to control population growth, but at women users with a choice of vaccines (shorter and longer acting, injectable and orally administered) to control their fertility. Others continued to design an instrument for population control (in India). However, this design line had to stop when the funding was discontinued (apparently in response to the pressure to do so mounted by women's health activists).

My own engagement in the controversy surrounding the antifertility vaccines was different than during the Norplant controversy. It is perhaps best characterized as *reflexive engagement*. In contrast to the Norplant case, I did not align myself, a priori, with one of the actors involved in the controversy. Instead, I reflected on the assumptions underlying the positions of both the women's health advocates and the scientists. I came to see that both parties were, in fact, heterogeneous, involving a variety of views and interests. Both referred to projected user needs to support their positions on what would be the best direction to go in designing this new technology. I engaged in reflexive dialogue with both parties. In the whole controversy, the voices of actual users of the technologies were absent.

CONSTRUCTIVE COLLABORATION: THE MICROBICIDE CASE

This realization led to a third form of engagement, that of constructive collaboration, in which I sought ways in which diverse groups of users of future technologies could be more closely involved in setting the parameters for the technology during the process of technology development.

The microbicide development process is an example of ways in which women's health advocates are involved in the early stages of the development of a new technology. In the late 1990s, in response to failures in promoting condoms to prevent HIV transmission, scientists with the support of some women's health advocates put forward proposals for the development of microbicides as alternative for women who are unable to negotiate safe-sex use of condoms with their partners. Microbicides are a wide range of chemical agents that are being developed to prevent/treat vaginal infections and sexually transmitted diseases including HIV. This plea for a new reproductive technology came at a time in which the longer-acting contraceptive technologies had become subject to heated debates concerning their gender scripts, as we have seen above. Researchers involved in microbicides development indicated a willingness to involve women in the research process, including the definition of criteria for safety, determination of research priorities, design and implementation of research protocols, and setting and monitoring of ethical standards and decisions on whether to pursue a fertility regulation method from one stage to the next. One of the vocal proponents of microbicides, Elias (a young male reproductive scientist) worked at the New York–based Population Council, the agency that had taken the lead role in developing Norplant. Elias developed the microbicide research program in close collaboration with Lori Heise, a vocal women's health activist.

The researchers at the Population Council decided to create an advisory body to bring women's health advocates' and users' concerns into the design of microbicides. A select group of internationally known women's health advocates was invited to act as advisers in a committee called WHAM (Women's Health Advocacy on Microbicides). I participated in this body as a women's health advocate and social scientist. In the latter capacity I was invited to propose methodological reforms in the clinical trial process that would enable users' views to be considered early in the development of the technology.

By setting up WHAM, the designers at the Population Council took the initiative to work with women's health advocates early in developing proposals for clinical trials to test the safety and efficacy of these new devices. The proposals emerging from the collaboration are unique: They suggest methodological reforms that allow experiential knowledge to feed into the biomedical routine of testing new technologies. Clinical testing of new technologies has been criticized for its lack of sensitivity to user experiences and its reliance on biomedical judgements on what is good and bad about a technology. The methodological reforms suggested in this case resulted in a design in which clinical trials become sites where safety, efficacy, and acceptability are negotiated between diverse groups of stakeholders, including medical researchers, end users, and women's health advocates.

Listening to users early on in product design can broaden the frameworks for clinical testing of new technologies, including attention to attributes of technologies that are likely to please or displease future users. Providing evidence to biomedical researchers and public health policy makers on user preferences is the most evident role that social scientists can play in the development of new reproductive technologies. It is indeed the most cited role in the body of literature on microbicides. I argued for a greater variety of roles of social scientists in the development of microbicides. Specifically, they should investigate particular issues at each stage of the development process.

1. When the idea of a new product is born, is it likely to benefit women? Is it needed? What alternatives are there? What are the risks? What is the social/power context in which it will be used? My counterparts at the Population Council did not consider these questions relevant; they considered themselves to have argued sufficiently that the products are needed. I suggested that one could argue that there is no need for microbicides, as male and female condoms are more effective in preventing transmission of HIV. HIV prevention programs should therefore promote condoms rather than less effective alternatives. On the other hand, I acknowledged that in situations where women cannot use condoms, microbicides (which can be used without men knowing) could be empowering, by allowing women to protect themselves against HIV.

2. What kinds of microbicides would women want? Based on multi-sited explorations on the use of vaginal products, I suggested that women want methods that protect against a wide range of reproductive tract infections (RTIs), including candida and bacterial vaginosis, which are very common. Also, women use vaginal products to increase sexual pleasure. And finally, some women may want to use microbicides to protect against HIV while intending to become pregnant. Protection against RTIs and enhancement of sexual pleasure, as well as spermicidal power, should therefore be considered as key design attributes.

3. When clinical trials are planned and conducted, what are users' safety and

efficacy concerns? I argued that user views and experiences need to be explored in the context of clinical trials, as this is the first point of contact between diverse groups of users and the new technologies. Such studies should include attention for the potentially empowering or oppressing role of the technology, in terms of gender-power relations and practices.

4. When the product is introduced on the market, in what way can users be empowered to make sensible health care choices? Building on past Norplant work, I called for attention for the introductory scripts of microbicides needed to ensure that they meet their feminist potential. I argued that microbicides should be offered in a context of choice, including male and female condoms, and informed choice out of these alternatives. I stressed that these systems issues need to be considered early in the design process.

I characterize my role in this controversy as one of constructive mediator. My role has been to reconceptualize clinical trials as sites for negotiations on safety, efficacy, and acceptability of new reproductive technologies and to add additional methods prior to and in conjunction with the trials, allowing for feedback from users of this new technology on perceived safety and efficacy, the effects of the technology on gender-power relations, and on sexuality and quality of life. The approach could be characterized as pragmatic. Researchers are involved in the development of new contraceptive technologies, and much can be gained from broadening the range of participants in defining what matters in the process. Issues of representation emerge; who can represent whom in the process? Issues of power are also highly relevant: whose values and interests count? In a pragmatic role of constructive mediation, such issues do not hinder engagement, but rather challenge the researcher to study the consequences of the reforms and ways to overcome problems.

CONCLUSION

In this chapter I have three aims: It described feminist concerns about the gender scripts of three new reproductive technologies, analysed how these concerns have shaped the (re)design of new reproductive technologies, and reflected on my own modes of engagement in the processes discussed. With respect to the first and second, we have seen how designers of three new reproductive technologies became more interested in involving users and representatives of users in setting parameters for and deciding on future directions in design. In the Norplant case, women's health advocates were concerned that the designers had compromised the health of users in favor of long-term efficacy. They were concerned about the disempowering effect of the contraceptive technology, which indeed proved to occur, as reflected in the problems that women encountered when demanding removal. The controversy led to guidelines on how to use the technology (introductory scripts) and not to changes in the hardware itself. The controversy concerning antifertility vaccines arose in an earlier stage of design. Vaccine prototypes were still being tested. Here the concerns of women's health advocates led to changes in gender scripts of the technologies: radically different kinds of vaccines were developed (shorter acting and different modes of administration). Researchers started emphasizing the "liberating" potentials of the technology rather than their population-control merits. In the third case history, the development of microbicides, we see that women's concerns were considered at an even earlier stage of development: before testing was conducted. The emphasis here was on developing methods for earlier and more systematic participation of women in the design processes.

With respect to the second aim, I have shown how my role as an activist-researcher shifted in these three cases from subaltern alignment (the Norplant case), to reflexive dialogue (the antifertility case), to constructive mediation (the microbicide case). These modes of engagement have in common that they require a balancing act between being a feminist activist and a social science researcher. In describing the shifts in approach, I do not intend to argue that the latter modes of engagement are better. All three have had an impact on the design of new reproductive technologies.

The first approach, subaltern alignment, is needed to give relatively powerless women a voice. This approach is possible once technologies are on the market, as I have shown in the Norplant case. However, such alignment should not simplify the issues. Views and needs of women differ. They are context dependent and historically contingent. In presenting women's views, researchers need to acknowledge this diversity. The second, reflexive dialogue, can help in unraveling fundamental differences in standpoints of the actors involved. In itself this approach does not have much power to change the gender scripts of technologies. It can, however, enable dialogue on contentious issues. Reflexivity led me to see more clearly the issue of representation. To whom do we grant power to speak on behalf of diverse users? Both reproductive scientists and women's health advocates were claiming to be legitimate spokespersons for future users. The problem in the antifertility controversy, the case in which I adopted this position, was that only a few users were being confronted with this technology; that is, only in the context of clinical trials. The third approach, constructive mediation, which I found myself adopting in the microbicide case, involves methods that enable researchers to explore diverse women's views and needs early in the development of new technologies. These methods allow for systematic consideration of women's views, behaviors, skills, and interests early in product development, thus limiting the risk that the technologies will turn out to be disempowering.

Through the three cases, I have been exploring how (future) users of preventive reproductive technologies can be involved in the delicate weighing of risks and benefits. A major constraint is that scientific information about Norplant, the antifertility vaccines, and any other new technology such as the microbicides is difficult to translate into language that is understood by people without a biomedical background. Materials need to be produced in simple language. But the agents involved in the translation process are likely to affect the translation, and messages received by the people can be oversimplified or biased. For example, researchers could state that to date no serious adverse effect of vaccines have been observed, omitting the theoretical risks of cross-reactions and autoimmune disease. Also, the way they have represented it to date as a "vaccine" could be considered misleading, as it indicates the positive image of vaccination against disease. Women's health advocates, on the other hand, have been criticized for putting too much emphasis on the risks of the anti-hCG vaccines, not explaining what the merits of the new technology could be.

I take as a point of departure that women make trade-offs between certain safety and efficacy characteristics of these reproductive technologies. It is crucial that the researchers try to understand what women want in their own terms, relating this to their explanatory models of fertility, infertility, and health and the realities of their lives (including attention to such issues as lack of autonomy in decisions concerning fertility, poverty, and ill health). Women's views and needs will, of course, differ across sociocultural settings.

Finally, I want to emphasize that there is a need to create space for researchers, women's health advocates, future users, and other actors in various sociocultural contexts to discuss what they see as valid research methods and appropriate requirements for and acceptable risks of new reproductive technologies. As I presented in the microbicide case, reviewing existing social science studies on fertility regulation, conducting community-based research, and reorienting controlled clinical trials can be a starting point for refocusing reproductive researchers toward the actual reproductive needs of people in different sociocultural settings. Ideally, such user-oriented design processes institutionalize a negotiation between actors on the way in which facts about the efficacy, safety, and acceptability of new reproductive technologies are investigated. This would give users space to help co-construct the framework through which scientists are designing and evaluating the new reproductive technologies.

References

Ada, G. L., and P. D. Griffin, eds. 1991. *Vaccines for Fertility Regulation: The Assessment of Their Safety and Efficacy.* Cambridge: Cambridge University Press.

Akhter, F. 1995. *Resisting Norplant: Women's Struggle in Bangladesh Against Coercion and Violence.* Dhaka: UBINIG.

Akrich, M. 1992. "The Description of Technical Objects." In W. Bijker and J. Law, eds., *Shaping Technology/Building Society: Studies in Sociotechnical Change,* 205–44. Cambridge, Mass.: MIT Press.

Anonymous. 1993. Declaration of the International Symposium on Contraceptive Research and Development for the Year 2000 and Beyond (March 8–10). Mexico City.

Bangladesh Fertility Research Program. 1981. "A New Birth Control Method: Norplant." Advertisement. *Holiday Weekly* (October 4).

Barzelatto, J. 1991. "Welcoming Address." In G. L. Ada and P. D. Griffin, eds., *Vaccines for Fertility Regulation: The Assessment of Their Safety and Efficacy.* Cambridge: Cambridge University Press.

Berer, M. 1997. "Why Reproductive Health and Rights: Because I Am a Woman." *Reproductive Health Matters* 5, no. 10: 16–20.

Berg, A., and M. Lie. 1993. "Feminism and Constructivism: Do Artefacts Have Gender?" *Science, Technology and Human Values* 20, no. 3: 332–51.

Bongaarts, J., and E. Johansson. 2000. *Future Trends in Contraception in the Developing World: Prevalence and Method Mix.* Population Council Report, no. 141. New York: Population Council.

Briggs, C. 1986. *Learning How to Ask: A Sociolinguistic Appraisal of the Role on the Interview in Social Science Research.* Cambridge: Cambridge University Press.

Correa, S. 1994. "Norplant in the Nineties: Realities, Dilemmas and Missing Pieces." In G. Sen and R. Snow, eds., *Power and Decision: The Social Control of Reproduction,* 287–311. Harvard Series on Population and International Health. Cambridge, Mass.: Harvard University Press.

Garcia-Moreno, C., and A. Claro. 1994. "Challenges for the Women's Health Movement: Women's Rights versus Population Control." In G. Sen, A. Germain, and L. C. Chen, eds., *Population Policies Reconsidered: Health, Empowerment, and Rights,* 47–63. Harvard Series on Population and International Health. Boston: Harvard School of Public Health.

Griffin, P. D. 1990. "Letters on Contraceptive Vaccine Research." *WGNRR Newsletter* 32: 6–7.

———, W. R. Jones, and V. C. Stevens. 1994. "Anti-Fertility Vaccines: Current Status and Implications for Family Planning Programmes." *Reproductive Health Matters* 2, no. 3: 108–13.

Gupta, S. K., and P. T. Koothan. 1990. "Relevance of Immuno-Contraceptive Vaccines for Population Control. I. Hormonal Immunocontraception." *Archivum Immunologiae et Therapiae Experimentalis (Warsz)* 38, no. 1–2: 47–60.

Haraway, D. J. 1991. *Simians, Cyborgs, and Women: The Reinvention of Nature.* London: Free Association Books.

Hardon, A. P. 1989. "An Analysis of Research on New Contraceptive hCG Vaccines." *WGNRR Newsletter* (April–June): 15.

———. 1992. "The Needs of Women versus the Interests of Family Planning Personnel, Policy Makers and Researchers: Conflicting Views on Safety and Acceptability of Contraceptives." *Social Science and Medicine* 35, no. 6: 753–66.

———. 1997. "Contesting Claims on the Safety and Acceptability of Anti-Fertility Vaccines." *Reproductive Health Matters* 5, no. 10: 68–82.

———. 2006. "Contraceptive Innovation: Reinventing the Script." *Social Science and Medicine* 62, no. 3: 614–27.

Hartmann, B. 1995. *Reproductive Rights and Wrongs* (revised ed.). Boston: South-End Press.

ICASC. 1980. "International Contraception, Abortion, and Sterilization Campaign: Women Decide." *ICASC Newsletter* 2.

IDRC. 1995. *Position Statement on Contraceptive Vaccine Research.* Ottawa: International Development Research Centre (IDRC), 2.

Jayaraman, K. S. 1997. "India Downgrades Family Planning Research." *Nature Medicine* 1, no. 5: 478.

Jones, W. R. 1982. *Immunological Fertility Regulation.* Oxford: Blackwell Scientific.

———, J. Bradley, and S. J. Judd. 1988. "Phase I Clinical Trial of a World Health Organization Birth Control Vaccine." *The Lancet* 1, no. 8598: 1295–98.

Kammen, J. van. 1999. "Representing Users' Bodies: The Gendered Development of Anti-Fertility Vaccines." *Science, Technology and Human Values* 23, no. 3: 307–37.

Keck, M. E., and K. A. Sikkink. 1998. *Activists beyond Borders: Advocacy Networks in International Politics.* Ithaca, N.Y.: Cornell University Press.

Macklin, R. 1994. "Combating the Potential for Abuse." *Reproductive Health Matters* 2, no. 4: 110–12.

Mintzes, B., A. P. Hardon, and J. Hanhart. 1993. *Norplant: Under Her Skin.* Delft: Eburon.

Mitchison, N. A. 1990. "Gonadotropin Vaccines." *Current Opinion in Immunology* 2, no. 5: 725–27.

Nash, H. A., et al. 1980. "Observations on the Antigenicity and Clinical Effects of a Candidate Anti-Pregnancy Vaccine: Beta-Subunit of Human Chorionic Gonadotropin Linked to Tetanus Toxoid." *Fertility and Sterility* 34: 328–35.

Oudshoorn, N. 1995. "Technologie en zorg: vriendinnen of vijanden." *Gezondheid, theorie in praktijk* 3: 278–89.

———, N., A. R. Saetman, and M. Lie. 2002. "On Gender and Things: Reflections on an Exhibition on Gendered Artifacts." *Women's Studies International Forum* 25, no. 4: 471–84.

Ravindran, S. T. K., and M. Berer. 1994. "Contraceptive Safety and Effectiveness: Re-Evaluating Women's Needs and Professional Criteria." *Reproductive Health Matters* 2, no. 3: 6–12.

Richter, J. 1992. "Research on Antifertility Vaccines Priority or Problem?" *WGNRR Newsletter* 39 (April–June): 13–18.

Schrater, A. F. 1992. "Contraceptive Vaccines: Promises and Problems." In Helen Bequaert Holmes, ed., *Issues in Reproductive Technology: An Anthology,* 31–52. New York: Garland.

———. 1994. "The Pros and Cons: Guarded Optimism." *Reproductive Health Matters* 2, no. 4: 189–10.

Schrijvers, J. 1991. "Dialectics of the Dialogical Ideal: Studying Down, Studying Sideways and Studying Up." In L. Nencel and P. Pels, eds., *Constructing Knowledge: Authority and Critique in Social Science,* 162–79. London: Sage.

Shervington, D. 1994. "Questioning the Wisdom in Changing a Part of the Whole." *Reproductive Health Matters* 2, no. 4: 108.

Talwar, G. P., O. Singh, and L.V. Rao. 1988. "An Improved Immunogen For Anti-Human Chorionic Gonadotropin Vaccine Eliciting Antibodies Reactive with a Conformation Native to the Hormone Without Cross-Reaction with Human Follicle Stimulating Hormone and Human Thyroid Stimulating Hormone." *Journal of Reproductive Immunology* 14, no. 3: 210.

———, et al. 1990. "Phase I Clinical Trials with Three Formulations of Anti-Human Chorionic Gonadotropin Vaccine." *Contraception* 41, no. 3: 301–16.

Viswanath, K., and P. Kirbat. 1997. *Genealogy of a Controversy: Development of Anti-Fertility Vaccines. Social Science and Immunization Country Study: India.* Working Paper No. 7. Delhi: Center for Development Economics, 16.

WGNRR. 1993a. "Call for a Stop on Research." *Women's Global Network on Reproductive Rights* (June).

———. 1993b. "Call for a Stop of Research on Anti-Fertility 'Vaccines.'" *Women's Global Network on Reproductive Rights* (November).

Woolgar, S. 1991. "Configuring the User: The Case of Usability Trials." In J. Law, ed., *A Sociology of Monsters: Essays on Power, Technology and Domination,* 58–100. London: Routledge.

World Health Organization (WHO). 1978. "Task Force on Immunological Methods for Fertility Regulation: Evaluating the Safety and Efficacy of Placental Antigen Vaccines for Fertility Regulation." *Clinical Experiment on Immunology* 33: 360–375.

———. 1990. *WHO Biennial Report on Human Reproduction.* Geneva: WHO.

———. 1991. *Creating Common Ground: Women's Perspectives on the Selection and Introduction of Fertility Regulating Technologies.* WHO/HRP/ITT/91. Geneva: WHO.

———. 1995a. *Technology Development and Assessment: Products under Development.* WHO Special Program on Research, Development, and Research Training in Human Reproduction. Geneva: WHO.

———. 1995b. *The Role of the Program in Technology Development and Assessment: A Discussion Paper* (October 1995). UNDP/UNFPA/World Bank/WHO Special Program on Research, Development and Research Training in Human Reproduction. Geneva: WHO.

Yanco, J., A. Will, and B. Stemerding. 1996. *Resistance on the Rise: Stop Anti-Fertility "Vaccines."* Amsterdam: WGNRR.

Letters

Beattie, K. 1993. Letter to R. Petchesky, April 30.

Benagiano, G. 1994. Letter to WGNRR, January 21, in response to the Call for a Stop to Research, signed by the Director of the Special Programme of Research, Development and Research Training in Human Reproduction.

Bezanson, K. A. 1997. Letter to K. Seabrooke, International Campaign against Population Control and Abusive Hazardous Contraceptives, January 24.

Catley-Carlson, M. 1994. Letter to B. Stemerding, June 28, signed as the President of the Population Council (with a request for wide distribution of the letter).

Gabelnick, H. L. 1994. Letter to B. Stemerding, March 2, signed as the Director of the CONRAD Program for Contraceptive Research and Development.

Hardon, A. 1993. Letter to K. Beattie, June 30.

Stemerding, B. 1996. Letter to all people involved in the campaign, January 24. Amsterdam: WGNRR.

7 Teaching Feminist Technology Design

FRANCES BRONET AND LINDA L. LAYNE

WE BEGIN WITH THE understanding that the way technologies are designed and built can "enhance the power, authority, and privilege of some over others" (Winner 1986). We recognize that in our society, "power, authority, and privilege" still fall disproportionately to men. "Artifacts have politics" (Winner 1986) and our goal is to change gender politics—to empower women (individually and collectively), to eliminate gender bias and to create a world with gender equity.

We recognize that women are not a homogenous group and that they have different needs and desires.[1] We also recognize that there are multiple forms of feminism (see Aengst and Layne, this volume) and that feminist designers will not come up with designs we can all agree on. But we believe that generating more, sustained, multivoiced debate about the feminist or antifeminist attributes of existing or emergent technologies will be of great benefit to the goal of improving women's lives.

We do not want to sit back and offer post facto critiques of new technologies but want to intervene proactively to influence design. One obvious way to do this is to equip the next generations of designers to work toward this goal.

Drawing on Bronet's sixteen years' experience of teaching interdisciplinary design, and Layne's experience over many of those years collaborating with her, we share our experience of trying to teach feminist technology design and lay out some suggestions for fostering feminism in design studios. We focus on two recent efforts on our part to teach the design of feminist technologies.

PRODUCT DESIGN AND INNOVATION STUDENTS, RENSSELAER, FALL 2006

In 1998 an undergraduate program in Product, Design, and Innovation (PDI) was initiated by John Schumacher, a philosopher in the Science and Technology Studies (STS) Department; Frances Bronet, a faculty member in the School of Architecture; and Gary Gabriele, a faculty member in the School of Engineering at Rensselaer. The curriculum combines the requirements of either mechanical engineering and STS and/or architecture and STS and is built around a design studio each semester that helps students learn how to create new products, environments, services, and media in the context of social needs and aesthetic and environmental concerns. Each studio teaches a combination of social science, technical and design skills, and methods (see Gabriele et al. 2002).

During the fall of 2006, third-year students (five female, sixteen male) took PDI Studio 5 with STS Professor Ron Eglash. Most of those students were simultaneously enrolled in my (Layne's) 400-level course, Gender, Science, and Technology, which was one of the courses that fulfilled their advanced STS option. In this course, students began with five weeks on gender and science followed by ten weeks on gender and technology, during which time their readings and class discussions addressed the following questions: Do artifacts have gender politics? What do we want of technology? How does change occur? Is technology gendered? If so, how? Can technology liberate women? Is technology inherently patriarchal? About ten weeks through the semester, I guest-lectured on feminist technology in Eglash's studio.

I began with Rensselaer's slogan, "Why not change the world?" and my own interpretation of that as a mandate to work toward making the world a better place for women. Because not all of the studio students were also taking Gender, Science, and Technology, I reviewed the argument laid out in Winner's "Do Artifacts Have Politics?" and then drew on Sue Rosser's (2006) description of twelve different types of feminism to give them a simplified set of working definitions.

1. Common sense: things or arrangements that benefit women are feminist. This would include artifacts and arrangements that acknowledge and/or enhance their power.
2. Liberal feminism: things that enhance or achieve equity. This version is premised on the idea that women need a fair share of the pie; the pie is fine as it is. Typical examples of this are found in calls for equal access to income, education, etc.
3. Radical feminism: based on the premise that we need to change the pie. The social world needs to be redesigned to suit women.
4. Essentialist feminism: a variation on radical feminism in that it also calls for changing the pie. In this version, women are inherently better than men and so if the world were to be redesigned according to women's inclinations, it would be more nurturing (less war, more devotion to humanitarian concerns, etc). I cautioned them that claims for "natural" or "essential" differences between men and women have historically been used to oppress women.

Next we discussed "technology," starting with the definition that technologies are artifacts and arrangements that "enhance the abilities of human beings." I discussed the difference between tools and technology, referring to the "technology = tools + knowledge and skill" formula and reminded them that technologies often have both multiple and unintended consequences.

I then asked them to work with a partner to delineate criteria to be used to evaluate whether or not something is a "feminist technology." The following are the criteria they generated:

1. Ergonomic: in discussion they realized there were two possible versions of this: (a) "fitting women rather than men" (gynocentric) or (b) as fitting women equally well as men"; i.e., a version of "universal design"(liberal)
2. More intuitive for users: by this they meant that it ought to be easy for women to figure out how to use a device
3. Enhances communication and social relationships, which they understood to be valued by women
4. Based on collaboration with women throughout the design process
5. Affordable to women, recognizing the current lack of financial equality between the sexes
6. Usable by women of all ages and generations
7. Of more benefit to women than men
8. Capable of counteracting androcentric technologies
9. The language used during the design process, inscribed on the artifact, and in the user manuals and advertising should be either (a) gender fair; i.e., equally applicable to women as men (liberal) or (b) gynocentric (radical).

I was very pleased with the exercise and share these criteria both as effective guides for use in the design and evaluation of new technologies, and as an indicator of how productive such an exercise can be.

Next I showed them a short (one minute, sixteen seconds) video clip from *Saturday Night Live* called "Kotex Classic" (http://hollyhockfarms.multiply.com/video/item/18/Kotex_Classic) in which attractive women are shown in backless evening gowns, bikinis, low slung pants, and tight-fitting clothing, wearing clearly visible sanitary napkins, with the bulk of the napkin showing through the fabric or the elastic belt exposed around a bare waist. The students moaned and howled in disgust and disbelief at these images as the video played. I then shared with them the insights of Vostral's research on how problematic tampons are as feminist technologies because they help women adjust to the male world by hiding the fact that they are menstruating and facilitating "passing" as a nonbleeder (2008,

Examples of sexism inscribed
on an artifact: Woman pictured
using baby-changing table on
Continental Airlines. Photo by
Linda Layne, 2008.

Men pictured on gym
equipment at RPI fitness center.
Photo credit: Ellen Esrock,
2008.

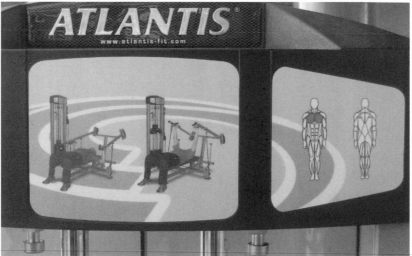

17–19). We compared the images in the video with the current fashion of having undergarments (both male and female) show, questioned how it happened that what used to be hidden is now made visible, addressed the right to dress for comfort and not be overly constrained by concern about how others might judge our appearance, and discussed the disproportionate extent to which women in our culture are so constrained.

One male student objected to the comparison of underwear with a pad, saying, "It is a sign of our civilization that we don't crap in public." We then discussed the double standard for men and women regarding public displays of bodily functions such as farting, burping, urinating, spitting, and vomiting, all of which are considered "unladylike" and all of which are engaged in publicly more often by men than by women.

During the last portion of class time, the students worked with their partners to see if they could think of any examples of "feminist technologies." This exercise was less successful than the generating of criteria but no less instructive. The technologies that they came up with were products that already existed, such as trucks and SUVs that had first been designed for men and lacked an ergonomic fit for women. Newer models have been adjusted to accommodate female users with adjustable steering wheel tilt, adjustable seats, and foot pedal placement for women. They also mentioned bicycles that are sized for both men and women and backpacks that fit women. With the exception of a team that suggested a pee-shoot that would allow women to pee while standing, all of the products they came up with had originally been designed for men.[2] In addition, as Eglash pointed out, none were products likely to be in high demand by the poor families this class was working with at the Ark (the Ark is a community-based after-school education center providing services that promote and reinforce positive development for at-risk youth in the City of Troy). Perhaps this assignment was not as successful because these students had not experienced much sexism or had not recognized it as such, and/or because the assignment was too broad. I may have had more success if I had provided them with a problem that faces women and asked them to imagine a technology that would address that.

A week or so later in Gender, Science, and Technology, Kate Boyer presented her work with Maia Boswell-Penc (this volume) investigating the breast pump as a feminist technology. In response to her suggestion that work spaces be redesigned to provide clean, private, attractive pumping rooms, one student objected that this would be a waste of resources, given how infrequently such a space would likely be used. I countered with the example of handicapped parking spaces that reserve the most valuable space in a parking lot for infrequent users. Further investigation shows, however, that although this was the case when these spaces were introduced, there has been growing use of and demand for these amenities[3] and one might reasonably expect this to be the case with nursing/pumping rooms. As they became more common, the use of and demand for them would grow and public awareness of these currently invisible forms of women's labor would increase.

AN NSF-FUNDED INTERDISCIPLINARY DESIGN WORKSHOP FOR FACULTY

University of Oregon, Winter 2007

Following a National Science Foundation (NSF) cross-disciplinary grant that helped launch Rensselaer's Product, Design, and Innovation curriculum, I (Bronet) co-ran[4] eight workshops on another four-year NSF grant (Design as a Creative Model for Technical Education) between 2003 and 2007. We initiated this project once we realized that although students were able to work in teams across disciplines, the faculty did not know how to work interdisciplinarily. Our experience had taught us that successful interdisciplinary experiences for the students require a faculty that is capable of interdisciplinary work. The best way to learn interdisciplinary design is to *do* it and so we developed a workshop for faculty.

The workshops were held at seven colleges or universities in the United States (including two workshops at a historically black college) and were attended by teams of university and college faculty and, on one occasion, K–12 teachers. They ranged from a half-day to five days but, regardless of their length, by the end of the workshop, teams had designed a product that was evaluated by external expert reviewers. Assignments included "design a device" (1) for kitchen waste and recycling, (2) that increased the fitness of unlikely exercisers, (3) that increased the likelihood of high school students going into engineering, (4) that encouraged interdisciplinarity among K–12 children. Results varied from the designs of political policies to physical, handheld products. Reviewers included city managers, architects, humanists, social scientists, and product designers.

After much discussion, Layne convinced me to try feminist technology as a topic. My concerns had been that the topic was too broad and ill-defined to get to a designed result in such a short period of time, and at the same time too prescriptive and narrow to allow the kind of creativity desired. Ultimately I decided on a more specific assignment: "Design a device that will increase gender equity in the home." This was given to ten University of Oregon faculty members and one patron of the school. They were divided into two teams with an effort to evenly distribute people with respect to gender, discipline, and rank. Team 1 had the associate department head of green chemistry, an associate professor of architecture, an associate professor of ceramics and products, the program director of interior architecture, the head of computer science, and another full professor computer scientist. Team 2 consisted of an associate professor of law, a cardiologist (the donor), a full professor of computer science, an assistant professor of digital arts, and an associate professor of architecture. Each team had three women. They were all volunteers.

The charrette, an intensive design process generally involving the collaboration of all project stakeholders to develop a proposal, unfolded as follows. The assignment "Design a device" was delivered. Each participant was asked to present what steps she or he would take if she or he followed her or his own discipline's norms to solve the assignment. Teams were then invited to come up with a collaborative process and a schedule in order to have a testable device at the end of the one and a half days. The teams were asked to follow the process they had established and to report back after a few hours. Team 1 considered several ideas, including eliminating chores altogether; establishing national policies; and creating a game called "kids rule the house," where children make decisions about the allotment of household chores with consideration of equity. Team 2 looked at product-placement models for the movie industry, giving awards to movies and advertisements if they reversed gender roles or portrayed men and women more equitably and in a way that made this arrangement seem ordinary. Both teams presented their work and discussions/critiques ensued.

Next, each team was asked to split into smaller work groups to pursue research, continue brainstorming, or start designing. After this breakout session, the ceramicist and green chemist who had been brainstorming together brought Team 1 a sketch of an artifact designed to make invisible household labor visible. It was a cast ceramic sphere that was split in two parts. If one member of a household did a task that might be invisible (e.g., scrubbing bathtubs), that person would lay out half of the sphere. When another household member recognized the work, he or she would put the missing half on the first piece. The team agreed to pursue this project.

Team 1 presented several versions of their technology: a craft-made object or set of objects that split into two—the invisible worker's sign and the acknowledger's sign, and a series of mass-produced objects (out of paper or other inexpensive material) to be sold in either specialty stores or big box outlets. Different designs/colors might be used by different family members. The hoped-for result was that by raising awareness of predicted inequalities, household labor patterns would change. Those doing less would do more, and everyone contributing would feel more fulfilled because their contribution would be acknowledged. One concern was that these objects might just add to household clutter and collect dust. Some noted that the "game" was in essence a "consciousness-raising device" and so envisioned periodic use. From time to time, when the more equitable division of household labor backslid, the intervention could be reintroduced. Some suggested thinking of it as a family game with a prize, like pizza night, once a certain target number of matches had been achieved. Others brought up that certain religions/cultures frowned on such recognition; for example, in Judaism it is seen as "self -indulgent" to seek credit (see http://www.targum.com/excerpts/goodman.html).

Team 2 created a not-for-profit foundation with a Web site that described their mission, their awards program, a rating system (comparable to the current parental advisory system), and small grant program for innovators in media productions that show equitable division of domestic rights and responsibilities. Although this product struck the reviewers as less original than Team 1's, the project was positively received because it was so thoroughly and professionally constructed, seemed eminently doable, and was likely to help increase gender equity in the home.

The iterative process followed during this workshop—design, presentation, design again with opportunities for critique, research, and presentation—is typical for group design practice. The feminist pedagogic innovations were the problem selection (feminist technology) and the methods used to make sure the participants (director and designers) stayed on task. I ensured this "vigilance" by selecting of reviewers who I knew would focus on the feminist elements of the design. Because this had been prearranged, I regularly reminded participants that the goal was to produce a feminist product or intervention. I realize now I could have augmented the feminist social agenda by passing out "expert documents" such as the seven questions Layne posed to her class, her class's nine criteria for feminist design, and Rosser's twelve types of feminism.

LESSONS FOR HOW TO TEACH THE DESIGN OF FEMINIST TECHNOLOGIES

Project Selection

Probably the most important lesson from these efforts concerns the most basic choice of project. We found that when assigned the task of designing to increase gender equity, faculty and students were able to do so. It was as simple as asking.

In the case of the NSF-funded workshop, the design problem was set for faculty, who perhaps understand gender bias better. In the case of the undergraduate course at Rensselaer, students were already several weeks into a course that focused on gender bias in technology. For most undergraduates, however, more assistance may be required. To the extent that students do not believe there is sexism, it will be hard for them to identify a problem that a feminist design could address. Instructors need to help students by providing information on existing gender-biased conditions.

The importance of project selection cannot be overstated. Our experience at Rensselaer and information gleaned from the Web sites of other engineering design programs indicates a strong bias for projects that are "male-valenced" or "male-friendly."[5] For example, in the 1990s, in one of the introductory, general manufacturing classes, Rensselaer engineering students were asked to machine a miniature cannon as part of teaching them how to use a lathe. Any number of objects could have been chosen for learning how to machine craft three-dimensionally.[6]

In their sophomore year at Rensselaer, students take Introduction to Engineering Design (IED), and for several years this course had as its project a device to "launch and aim." In fall 2003, the IED students were asked to design a machine that could compete at a carnival game that involved tossing or launching an object onto a playing field. One of the professors, Bill Foley, recalls, "We allowed dart launching to balloons . . . and had students using visual systems to acquire targets and aim the launcher. One launcher was able to launch darts through concrete blocks using nail gun propulsion methods." As recently as fall 2004, the project was to design, develop, fabricate, test, and compete with a computer-directed robot that sends balls through targets on a course to score points. According to the syllabus, this project choice was inspired by the release of the movie *Dodge-ball* (a 2004 guy-humor flick), the U.S. FIRST competition in which remotely controlled machines engage in direct face-to-face competition on the playing field, and a March 2004 Defense Department–sponsored Mojave Desert road race competition between teams ranging from high school students to defense contractors that involved designing a robot vehicle that could travel off-road 150 miles across the desert in ten hours or less. In spring 2004, the goal was to design and build a machine that could launch small balsa wood model gliders to set locations in a flight area. Rosser (1995, 5) suggests that in order to make physical science, math, and engineering more appealing to women, one should "undertake fewer experiments likely to have applications of direct benefit to the military and propose more experiments to explore problems of social concern," adding that many girls and young women are uncomfortable engaging in experiments that . . . seem useful only for calculating a rocket or bomb trajectory."

This type of project choice is by no means unique to Rensselaer. Fifteen of the sixteen student design projects described for UC Davis in 2008 were male-valenced, including a radio-controlled airplane, a remote-controlled underwater vehicle, six car projects (including the same Defense Department–funded desert race in which Rensselaer students participated),[7] two robot projects, a concrete canoe, and a steel bridge. Their "Design Clinic," which touts a "wide range of industrial design projects," and in theory might include projects like breast pumps, gives as examples "human-powered or hybrid-electric vehicles, rocket engine clusters, and Caltrans maintenance trucks"; that is, the very same types of male-friendly projects covered in the other courses (American Society for Engineering Education 2008). The only project or course for which a woman is listed as the faculty adviser is the environmental engineering one, and its title is "Water Treatment from Your Kitchen . . . and Beyond," a competition in which students "design and construct a water treatment system that will contain, treat, and discharge a sample of water contaminated with organic and inorganic matter."[8]

Most of these projects are, in fact, "competitions" sponsored by engineering societies (e.g., American Society of Civil Engineers, Society for Aeronautical Engineers, Society for Automotive Engineers), government agencies (DARPA, the U.S. Department of Energy), and/or corporations such as Lockheed Martin or General Motors. Rosser suggests that one way to attract girls and young women to engineering is to use "more cooperative rather than competitive pedagogical methods. Although male students may thrive on competing to see who can finish the problem first, females prefer and perform better in situations in which everyone wins" (1995, 11).

If we make the empowerment of women and increasing gender equity explicit goals of design studio projects, the likelihood of feminist technologies being designed will increase. This practice will also likely attract more women into design fields. Studies of barriers to women joining science, technology, engineering and mathematics (STEM) consistently find that women are attracted to fields where they feel the work is meaningful. In their study of high school girls who are "academically prepared for college-level STEM courses," the Extraordinary Women Engineers Project (2005), a national initiative to encourage girls to consider pursuing a degree and subsequent career in engineering, found girls want careers that are "relevant," where they will "be able to make a difference" (Extraordinary Women Engineers 2005, 3).[9]

The kinds of projects with which engineering is typically associated do not appeal to most young women. In response to the question, "What are the first two words that come to mind when you hear 'engineer'?" the girls' answers included "men, cars, engines, trains, bridges, boring, boys" (Extraordinary Women Engineers 2005, 9), words that, in fact, accurately describe many of the design projects just outlined. The report also notes that one of the barriers to interest among high school girls is "they do not understand how engineering helps people" (Extraordinary Women Engineers 2005, 11). This is one of the reasons that the Engineering Projects in Community Service (EPICS) program at Purdue University has had such success in attracting female engineering and computer science students. Compared with the 10 percent of women majoring in electrical and computer engineering (ECE) and 12 percent in mechanical engineering (ME), 20 percent of ECE and ME students in EPICS were women, and the enrollment rate in EPICS for women in computer science (CS) was nearly three times the enrollment rate of undergraduate women in CS. Women have also taken on team-leader roles at unusually high rates in this program (30 percent of such roles during a time when women represented 20 percent of the EPICS students).

Thirty-six multidisciplinary projects are described on their Web site, including educational materials for the Community Park Zoo, designing and building a stage for a Community and Family Resource Center, several water-quality projects, and several assistive technology projects for community organizations and the dean of students office. Some of these have a mechanical engineering focus and others have an electrical computing engineering focus; several projects involve retirement communities, new designs for more energy-efficient Habitat for Humanity houses, science and technology exhibits for the children's museum, a redesigned gym for the YWCA, the design of a new soccer complex for youth soccer, an Interactive Rainforest Room at Happy Hollow Elementary School, and software to spark the interest of girls in technology in partnership with the Anita Gorg Institute for Women and Technology in Palo Alto (Purdue University 2008).

A similar success was achieved at Rensselaer with a hybrid Biomedical Capstone Design and Architecture Design course that sought to improve the design of neonatal intensive care units (NICU). Five of the six architecture students who selected this course were women and six out of six of the biomedical engineering students were, as were the two primary instructors, Bronet and Natacha DePaola, chair of biomedical engineering.

Keeping the Feminist Agenda at the Forefront of the Course

In addition to the explicit project criteria presented at the beginning, it is important to keep the feminist agenda present throughout the course. In studio courses, students meet a set number of times a week (two to four times) for extended periods (three to seven hours), where they engage with their faculty in ratios of about one to fifteen. There are a number of parallel methods involved in the studio process: the desk critique, where the faculty member and student meet at the student's table or computer to review the work on a one-to-one basis; the brief, "just-in-time" lecture; the pin-up, where the faculty member and, perhaps, guest meet with a group of students working on the same design problem to review the work at an interim point; the review, usually at the midterm and end of a project involving all the students and invited guest critics. Students must get feedback on their design in terms of the goal of increasing gender equity at each of these stages.

The two keys for ensuring this are (1) instructors and (2) reviewers. If the course is team-taught, it is essential that the majority of the instructors be committed to the feminist agenda. If there is just a single "feminist expert" instructor, her or his voice is likely to get lost. A feminist agenda is a social agenda and prior experience with the PDI Studio 1 at Rensselaer taught us that no matter how much the instructors plan on "integrating the social" into a design studio, once the course gets under way, the "social" agenda tends to get pushed aside. Although the PDI program was established explicitly to integrate engineering, aesthetic, and social agendas, in this interdisciplinary design studio, the social had a tendency to fall away during the course. There were multiple reasons for this: (a) Studio 1 was taught as an offshoot of the introductory architecture design studio, (b) the instructors consisted of four architects, one social scientist, and one engineer, (c) it took place in the architecture studio space, (d) the PDI students were mixed in with the more numerous first-year architectural students, (e) the social science faculty member was in studio only during the scheduled class hours and did not return to the studio other times during the week, such as evenings and weekends, as did the architecture instructors, and (f) because the studio was a design/build studio, during the rush to complete the projects, other matters took precedence. Each of these structural elements diminished the likelihood that the goal of integrating a social agenda into the design process would succeed. All but the last of these is avoidable and should be guarded against when striving to integrate feminism into a design studio.

Whether a design studio is team- or individually taught, reviewer selection is a key factor in ensuring that a feminist agenda be sustained. During a semester-long course, student work is typically reviewed by guest reviewers at midterm and upon completion at the end of term. If the course is individually taught, the instructor will typically also bring in outside experts for "just-in-time" lectures and the pin-ups. It is important to make sure guest reviewers understand and support the feminist design agenda.

The social cannot be the only criteria by which student work is judged, however, because if the technical and aesthetic dimensions of the design are not excellent, the student will not find success in the design world. In design studios, often if the technical and aesthetic dimensions of a project are excellent, the project is considered a success. A project rarely fails if the social element of the design is not good or evident; it would if the formal and technical elements are subpar. Because we want feminist design to be the priority, a project will need to fail if feminist understanding is missing. The social, in this case, feminist, agenda must be held equally essential to the other more standard measures of design success.

It is important that students who explicitly embrace feminist design are not discouraged. For example, Zara Logue, a female product design student at California College of Arts and Crafts and amateur race-car driver, chose as her thesis project women in the world of sport compact car modification and amateur racing. Based on ethnographic research, she learned that the fire suits that must be worn for safety are ill-suited to women. She felt they made her look like "the Marshmallow Man from *Ghostbusters*" and, because of the fire-resistant fabric, they were uncomfortably warm. Male racers coped with this by unzipping and taking off the top of the suit (much as male surfers do with full wet suits). Furthermore, "you are required to drink fluids whenever you are off the track in order to stay hydrated, which leads to many bathroom breaks." As part of her final project, she designed an alternative, the Eve racing suit, which featured a zipper to allow women to use the restroom without removing the entire suit, "pit zips" to allow cooling the underarms, and, in one version, openings for breast-feeding (Logue 2008).

The reviewers were older male product designers who were turned off by what they saw as a "girl power" project (as were some of her male peers) and were discouraging about the potential market for the product. They preferred the performance art element of the project in which she designed a two-piece bathing suit out of car wash mitts, and in a parody of a "bikini car wash" (there are many videos under this rubric on YouTube of young women dressed in bikinis washing cars), filmed herself using this artifact to wash her car. Reviewers can be expected to do as these "men in black" apparently did; that is, to default to what they know best, the artistic, the formal, and the aesthetic. This example illustrates the need for feminists to be well represented on review panels.

Curriculum Development

Feminism needs to be integrated throughout the design curriculum, not restricted to a single project or even a special course but pervasive, just as green building has become embedded in the curricula of the University of Oregon architectural design program. At the University of Oregon, green design is now nearly as ordinary as designing living rooms with windows. We want to make the goal of increasing gender equity so common that it is taken into consideration whenever design is undertaken.

This can happen in a number of ways. A decision emerging from junior and/or senior faculty and supported by the chair or head of a department can create a program that makes feminist design a priority. This is what has occurred in many design schools with respect to green design. The adoption of feminist design as a priority might also emerge from student demand and/or from an administrative mandate. Schools that have been successful at green design, responsible (including environmental, social, and economic justice) design, or community design have done this through strategies that pervade the curriculum. There are either an array of specific courses and studios that make this an explicit goal, or the particular social agenda is embedded as ordinary in every part of the program. Because the studio is the focus of design curricula across North America and perhaps much of the world, a feminist focus must penetrate studio courses.

Changing Studio Design Culture

This brings us to the issue of how the design culture inculcated in studio courses at universities is biased against women. In their book *Women's Science: Learning and Succeeding from the Margins,* Eisenhart and Finkel found that women tended to cluster in areas of science that required "more flexible commitments of time, space, and professional identity than the 'greedy' . . . elite science[s]," which required "more of one's time, tighter constraints on appropriate workplaces, and narrower identities and networks of power" (1998, 229). This distinction is probably confronted first in college where students recognize "greedy versus more flexible courses [and] degree programs." The alternative, lower-status sites they studied, like environmental biology, "made high demands, too, but not such extraordinary ones and in ways so likely to disadvantage young women" (1998, 230).

The design studio environment is a "greedy" one; one that reveres "all-nighters" and working around the clock on the design project to the detriment of all other courses, relationships, and activities. The expectation that students will put in forty-plus hours per week in studio courses alone is one element of the macho culture it venerates and is established as a baseline for behaviors in professional practice. Designers ultimately are poorly paid per hour, because the expectation is that overtime is normal, and it discourages life outside the studio or office.

The Extraordinary Women Engineers Project (2005) found girls want careers that will "be flexible" (2005, 3). "Students are looking for a job that allows time for family, hobbies, and travel. High school students who were academically qualified for college math and sciences classes explained their lack of interest saying, 'My career can't consume all of my time . . . I need free time to do a lot of other things I want to accomplish before I die, and when I have a family, I need time to spend with them'" (Extraordinary Women Engineers 2005, 13) Of engineering, one observed, "There seems to be less flexibility for those women that want to have a family" (Extraordinary Women Engineers 2005, 12). The same could be said of industrial and architectural design.

One intervention being discussed at University of Oregon is to pilot a studio where the number of hours students spend on work in the studio is limited, as an effort to challenge macho culture and to see if the quality of work is compromised. The proposal is called 37.5 and is intended to make design culture more woman/family friendly and, in so doing, to even the playing field.

CONCLUSIONS

In the United States, one of the primary ways the feminist movement has attempted to overcome gender bias has been through changes in the law and public policy; for example, *Roe v. Wade* or the Equal Rights Amendment. Technologies are "forms of legislation" (Winner 1977) that have not yet received the level of feminist attention they deserve. As Winner (2002, 1) observed, "Just as surely as . . . the laws and regulations of government, technological design is a place where some basic decisions are made about the identities and relationships, power and status, life chances and limits upon these chances."[10]

The disability rights movement understood this and demanded a "thoroughly reengineered world." It is time for feminists to do the same. We, too, have the right to shape and reform technologies because "the way they are structured, how they operate, what conditions and requirements they impose" profoundly affect women's possibilities for action and fulfillment (Winner 2005, 6).[11]

For several decades, feminist scholars have fruitfully asked "do artifacts have gender?" It is time now to turn our focus to figuring out "how we can design artifacts to change gender" (adapted from Winner 2002, 7). An obvious place to begin is in the education of young designers. Some of the things we can do to facilitate the adoption of a feminist design agenda would be to (1) build a set of exemplary projects that could be disseminated; (2) create a listserv of feminist design critics who can address the feminist as well as the technical and aesthetic elements of a design; (3) hold meetings and workshops at conferences to reinforce current and inspirational feminist work; (4) produce a list of potential industry, not-for-profit, and government funding sponsors that would support feminist design projects in product, engineering, and systems design courses, (5) have professional organizations like the American Society of Engineering Educators, Association of Collegiate Schools of Architecture, or Industrial Designers Society of America offer prizes for the best feminist design.

It is encouraging that at a growing number of experimental and cutting-edge architecture and design schools, we are beginning to see a greater commitment to social engagement in the design enterprise.[12] Let's make sure that when schools of design embrace "social agendas," feminism has a central place.

NOTES

We are grateful to the National Science Foundation for supporting the development of the PDI program and funding the workshop described in this chapter. Thanks to Ron Eglash for the opportunity to guest lecture to his students and for providing notes on what transpired.

1. We recognize that gender bias is only one form of social injustice that new technologies could/should address and that there is overlap between women's needs and those of other groups, such as people with disabilities and seniors. In the United States, more women are disabled than men—28.6 million to 25.3 million (Jans and Stoddard 1999) and more of the population over sixty-five are women (in this age range, "women outnumbered men by 5.3 million") (Spraggins 2003, 2).

2. "Taking a design formerly associated with one gender and transforming it so that their features are more compatible with another gender" is one of the two design strategies that Winner identifies as having been used to accommodate women's needs. The other is "to make useful instruments gender neutral" (2002, 3).

3. The number of people using these spots (both legally and illegally) has led to stricter criteria in the issuing of handicapped permits and a vigilante movement to stop handicapped parking fraud (Handicapped Fraud Organization 2008).

4. The first four workshops were coorganized with Gary Gabriele.

5. This is an adaptation of Rosser's (1997) concept "Female-Friendly Science."

6. Wikipedia notes: "Examples of objects that can be produced on a lathe include candlestick holders, cue sticks, table legs, bowls, baseball bats, crankshafts and camshafts."

7. A hidden constraint on all these projects is the funding source; many of these courses and competitions are supported by the military-industrial complex. As universities become less state-supported, will such male-valenced projects become even more pervasive?

8. The largest concentration of women engineering majors in 2007 was environmental engineering with 44.5 percent of bachelor's degrees awarded to women (American Society for Engineering Education 2008). The next most popular areas are biomedical (38.2 percent), chemical (36.3 percent), biological and agricultural engineering (32.6 percent), and industrial/manufacturing (30 percent) (American Society for Engineering Education 2008).

9. WGBH Educational Foundation conducted qualitative research from June 2004 to January 2005 on high school girls, science and math teachers, school counselors, male and

female college engineering students, and engineers consisting of six focus groups (four in person; two online), and five online surveys (Extraordinary Women Engineers 2005).

10. The example he gives is the placement of baby-changing stations in men's bathrooms and his reaction the first time he saw one in the early 1990s—he thought he must have mistakenly entered the women's restroom.

11. Although we recognize that artifacts have politics (Winner 1986), that physical characteristics and qualities can be valenced toward certain "forms of life" and political arrangements, we also recognize that even if feminist gender scripts are inscribed into a technology, the way an artifact will be used and understood will be largely influenced by the patriarchal status quo (Woodhouse and Sarewitz 2007). In other words, we recognize both the importance of engaging in change at this level and also the limitations of approaching change in this way.

12. One example is Syracuse University's School of Architecture's UPSTATE program. Inspired by Chancellor Nancy Cantor's "scholarship in action" principles, this interdisciplinary center for design was founded in 2005 to "apply innovative, experimental design research to challenges faced by real-world communities" (Syracuse University 2009). Southern California Institute of Architecture lists community outreach as one of its core values.

References

American Society for Engineering Education. 2008. Percentage of Bachelor's Degrees Awarded to Women by Discipline (2007) *Connections* (July). www.connections@asee .org.
———. http://www.asee.org/publications/profiles/colleges/. (Accessed May 2008.)
Eisenhart, Margaret A., and Elizabeth Finkel. 1998. *Women's Science: Learning and Succeeding from the Margins.* Chicago: University of Chicago Press.
Extraordinary Women Engineers. 2005. Final Report (April). www.engineeringwomen .org.
Gabriele, Gary A., Frances Bronet, Larry Kagen, Ron Eglash, Jeff Hannigan, David Hess, and Barbara Seruya. 2002. Product Design and Innovation: A New Curriculum Combining the Social Science. Design International Conference on Engineering Education (August 18–21). Manchester, UK.
Handicapped Fraud Organization. 2008. www.handicappedfraud.org. (Accessed May 2008.)
Jans, Lita, and Susan Stoddard. 1999. Chartbook on Women and Disability in the United States. Washington, D.C.: U.S. Department of Education, National Institute on Disability and Rehabilitation Research. H133D50017–96.
Logue, Zara. 2008. www.zaralogue.com/portfolio. (Accessed May 2008.)
Purdue University Engineering Projects in Community Service. 2008. http://epics.ecn. purdue.edu/projects/index.php. (Accessed May 2008.)
Rosser, Sue V., ed. 1995. *Teaching the Majority: Breaking the Gender Barrier in Science, Mathematics, and Engineering.* New York: Teachers College Press.

———. 1997. *Re-Engineering Female-Friendly Science.* New York: Teachers College Press.

———. 2006. "Using the Lenses of Feminist Theories to Focus on Women and Technology." In Mary Frank Fox, Deborah G. Johnson, and Sue Rosser, eds., *Women, Gender, and Technology,* 13–46. Urbana: University of Illinois Press.

Spraggins, Renee E. 2003. "Women and Men in the United States: March 2002." *Current Population Reports.* Washington, D.C.: U.S. Census Bureau.

Syracuse University. 2009. School of Architecture. http://soa.syr.edu/. (Accessed May 2009.)

Winner, Langdon. 1977. "Technology as Legislation." In *Autonomous Technology: Technics-out-of-Control as a Theme in Political Thought,* 317–25. Cambridge, Mass.: MIT Press.

———. 1986. "Do Artifacts Have Politics?" In *The Whale and Reactor: A Search for Limits in an Age of High Technology,* 19–49. Chicago: University of Chicago Press.

———. 2002. Gender Politics and Technological Design. Paper delivered at the Women's Studies Center, Colgate University, October 22.

———. 2005. Is There a Right to Shape Technology? Presented at Bard College, December 15.

Woodhouse, Edward, and Daniel Sarewitz. 2007. "Science Policies for Reducing Social Inequities." *Science and Public Policy* 34, no. 3: 139–50.

8   A Feminist Inventor's Studio

SHARRA L. VOSTRAL AND DEANA MCDONAGH

WHILE A FACULTY MEMBER in the Science and Technology Studies Department at Rensselaer Polytechnic Institute, I, Sharra L. Vostral, took turns teaching the course Gender, Science and Technology with Linda L. Layne and Nancy Campbell. Most of the students enrolled in that class were engineering majors, and I learned that hands-on assignments in which I asked them to design and even fabricate reimagined gendered objects helped them put feminist theory into practice. After beginning work on this volume and moving to the University of Illinois at Urbana-Champaign, where I hold a joint appointment in gender and women's studies and history, I realized the environment was ripe to put gendered design into action, and I proposed a feminist inventor's studio.

Like Rensselaer, the University of Illinois has a strong tradition of educating scientists and engineers. Unlike Rensselaer, the majority of students are admitted to the College of Liberal Arts and Sciences, and the college supports an active program in gender and women's studies. As adept as the Rensselaer students were in managing technology and submitting full-blown AutoCAD drawings, they were under-read in feminist theory. In contrast, the University of Illinois students often displayed hesitation about technology, yet had no qualms about calling out another student on his or her heteronormative or homonormative assumptions. It occurred to me that I needed to get these populations talking, and a feminist inventor's studio that brought both groups together would encourage interdisciplinary conversation and provide powerful tools of analysis for different types of students.

I also realized that in order to give the class further legitimacy, I needed to co-teach it with someone trained in design, and I needed institutional support. Through various introductions, I met Deana McDonagh, who holds a position in the School of Art and Design and the Beckman Institute for Advanced Science and Technology at the University of Illinois. Her enthusiasm for the idea let me know I had found the right person with whom to co-teach the studio. Her expertise as an industrial designer with special training in emotional responses to objects, combined with my training in women's relationships to science and technology, created a promising team to bring woman-centered design pedagogy to the college classroom. We also earned a faculty fellowship through the Academy for Entrepreneurial Leadership, funded by the Kauffman Foundation, to design the course, support students with materials, and pay for a teaching assistant. The combination of business, design, and focus on women-as-users brought together different schools at the university in a unique course setting. In addition, it was an opportune time to promote feminist technology and invention given the current agenda at Illinois to support new and emerging technologies.

As a matter of reference, gender and women's studies is couched among a broad network of faculty members from across the university, involving a range of academic projects, which include strong links to other units on campus to provide a rich intellectual environment for research, course work, and associated activities. In addition, because of the development of iFoundry, the Illinois Foundry for Tech Vision and Leadership, a unit of the College of Engineering, the relationship with that school is strengthening. iFoundry aims to rejuvenate engineering education by better incorporating the arts, humanities, and social sciences to inspire creativity and innovation, and both McDonagh and I have been appointed as fellows to this new initiative. As instructors with multiple affiliations, we are poised to place gender at the center of technological initiatives, moving beyond interpretation and post facto critiques to proactive feminist shaping of research and innovation.

McDonagh and I approached the course with the understanding that gender shapes not only understandings of people and spaces, but also products and objects. In addition, we believed that shifts in attitudes about gender could inspire new products for the market, and, conversely, that the new deployment of existing products had the potential to alter our understandings of gender. For example, during the early 1980s, my colleague Langdon Winner walked into the restroom in an upstate New York airport. As he put it, "I turned on my heel" and walked out the door, because he thought he had entered the women's side. He checked the door, which indeed posted the universal symbol for the men's room, and tentatively walked in again, slightly baffled. The object that challenged his assumption of the men's room and indicated he was mistakenly in the women's room was the installation of a diaper-changing table. At this time, there was no expectation that men would need this technology because that messy work was presumably left to women. Yet, the installation of a diaper-changing table holds significant political implications, along with the new expectation that men should indeed make use of it. Technological objects possess one means to remedy social and political disadvantages, so that designing and using objects specifically meant to disrupt inequality between men and women can promote women's interests.

As women and men increasingly grapple with the gendered aspects of design, work, and home life, products—whether changing tables in men's rooms or more convenient forms of birth control or simply seat belts that work on differently sized people—that creatively and insightfully take gender into account can catalyze societal change and be inspired by entrepreneurial approaches.

The Feminist Inventor's Studio was premised on a fundamental notion: design that places gender equity at the center of the creative process produces technologies and products that improve women's lives. It is clear that environmentalism has had great success in pressing for sustainable design, and most architecture programs have now embedded these tenets so that they are unquestioned assumptions in the design process. However, this is not the case for gender. To make this point, it is informative to look at one of the standard reference works for industrial designers, *The Measure of Man,* compiled by Henry Dreyfuss and first published in 1960 and updated shortly thereafter in 1967. At the time, the book transformed the industry because one collection contained "a miniature 'encyclopedia' of human factors data for the industrial designer, presented in graphic form" (Dreyfuss 1960, 4). This anthropometric data provided extensive measurements, charts, and numerical values for human-centered design. Although women were included, they would not appear in the title until 1993. In the twenty-six year period between the updates, which witnessed the rise of the feminist movement, the book remained stagnant. For example, in the image depicting a bicycle rider, only a man rode. Human strength charts equated a strong woman with an average man. Thus, men remained centered as the norm, with women being measured in comparison to them, thus privileging design toward the male body. This omission becomes all the more costly for producers because female consumers and managers of household economies increasingly outnumber males in the same categories.

As we prepare ourselves for demographic shifts of all sorts—an aging population on a global scale and ethnic diversity among women, for example—designers need to ensure that products and technologies are presented to the user that are more intuitive, more appropriate, and satisfy needs beyond the functional. With this approach, design is less about generating more products for the market and more about creating positive user experiences. These objects may be experienced directly by women or be designed for men to facilitate changes in work parity. By addressing how men and women use objects in terms of gender and gender expectations, new entrepreneurial avenues can open in design innovations.

The course also offered an important means to address technological literacy and competency in liberal arts and sciences students. Although there are many former engineers and scientists that have found their way to gender and women's studies classes, the Feminist Inventor's Studio was not intended just for them. The course encouraged humanities students, who do not necessarily identify themselves as technology focused, to apply gendered critique to a problem and convert that into action. It promoted their technological confidence by introducing it through a topic for which they are already passionate. By taking their sense of social fairness, equity, and democracy and applying it to design woman-friendly objects, they had the opportunity to gain skills in electronics, AutoCAD, and fabrication, for example. As liberal arts and sciences students, access to this type of training is limited at best. By using entrepreneurial approaches, this class encouraged critical thinking skills vaunted in humanities to solve a gendered design problem. It challenged design students, who are experts at reading a visual culture, to use language, terminology, and theory to analyze and identify gender-in-motion. Gender and women's studies and science and technology studies provide useful handles to clarify any given question, action, or phenomenon. By familiarizing themselves with language and concepts of gender, students became instruments of change and enabled the co-creation of knowledge between teammates.

The course was divided into three five-week sections: theory, field research, and design generation/repurposing. The first part introduced students to feminist theory, gender and design, and foundational elements to critically engage objects. Students were introduced to the strong legacy of women becoming entrepreneurs as a result of the inadequacy of a product, to provide both inspiration and also pragmatic wisdom about solving technical problems (one example is Susie Hewson, who founded Natracare, as discussed in chapter 5). The second part of the class asked students to identify a problem, find an opportunity, and think about solutions. An assignment that prompted students to think about design possibilities for women users was "Redesigning the Speculum." Most adult women are familiar with the medical device used by gynecologists to prop open the vaginal walls during a pelvic exam. Most will also tell you that this examination is not particularly comfortable, in part because of the construction of the steel or plastic speculum itself. A quick patent search reveals that the speculum has not changed in shape or style much since the one designed in 1892. The assignment asked students to use the lens of gender to ask how might we make a better speculum or, better yet, find ways to abandon it altogether. For example, what if there were a way to embed sensors in it to detect pathogens? Or use a material that was warm to the touch? In groups, students brainstormed a variety of solutions. As instructors, we demanded on a daily basis that students design for women; otherwise, we found that many too easily unconsciously reverted to the "male-as-norm" mode of thinking.

The final portion of the class required students to team up and generate a model, which became the focus of a final presentation with story boards and PowerPoint and was evaluated by an outside panel. This panel was composed of an affiliate member from gender and women's studies, art and design, and the School of Business, and each member was asked to rate the success of the projects based on the aesthetics and functionality of the project and how well it met feminist design parameters. Students also received a small stipend of seventy-five dollars, funded through the grant, to help defray the cost of materials for the prototype.

As students learned, careful consideration of the gendered aspects of design did not displace aesthetics and functionality, and indeed often enhanced them. Teams came up with ideas such as a waste container for used tampons for women on camping trips; an adjustable tool belt for housecleaning utensils; and a closet rack system to hang bras and undergarments. As instructors we asked and required students to include gender as a design element, with their success assessed and reflected in their project grades. Linda Layne and Frances Bronet (this volume) found that constructing class objectives to privilege feminist interventions, structuring class projects that value social networks and community application, and rewarding woman-friendly designs were effective pedagogic practices, and ones that were followed in the Feminist Inventor's Studio.

Because only the designers were trained to use prototype equipment, we waited with bated breath for the opening of the proposed Fab Lab in the spring of 2009. The Fab Lab fabrication laboratory is a small-scale workshop open to the community with sponsorship ties to MIT's Center for Bits and Atoms. The Fab Lab "program has strong connections with the technical outreach activities of a number of partner organizations, around the emerging possibility for ordinary people to not just learn about science and engineering but actually design machines and make measurements that are relevant to improving the quality of their lives" (Fab Lab). The Fab Lab was an ideal locale to promote invention, collaboration, and cooperative work and encourage entrepreneurial methods to redress gender inequities based on technological designs. The Fab Lab makes it possible for us to envision alternative ways of teaching because of the new resources at hand and provides the ability for humanities students to be trained and use rapid prototype machines. Because of delays, we were not able to utilize the Fab Lab, but intend to incorporate it into future courses.

In an age where sustainable design is not only a buzzword but a common practice, and democratic design is a movement to consciously design with many constituents in mind, now is the time to capitalize on feminist design by incorporating gender as an element of any project's critique. Why not promote more broad-based thinking to prevent harm to women, and in fact include pro-woman elements in the design process from the very start? The Feminist Inventor's Studio served as a conduit for humanities, tinkering, and entrepreneurialism to come together in a course, allowing non-design majors to practice creative problem solving, and non-humanities majors to practice critical gender inquiry in the social arena.

References

Dreyfuss, Henry. 1960. *The Measure of Man: Human Factors in Design.* New York: Whitney Library of Design.

Fab Lab. N.d. MIT's Center for Bits and Atoms. Electronic document. http://fab.cba.mit .edu/. (Accessed August 8, 2008.)

9 What We Now Know about Feminist Technologies

SHIRLEY GORENSTEIN

PRODUCING FEMINIST TECHNOLOGIES would seem to be simply a matter of combining the goals of feminism with the practicality of technology. However, in attempting to join the two, the authors of this book found themselves confronted by unexpectedly puzzling questions. Facing those conundrums required them to not only revisit concepts of feminism and technology, but also to create de novo the concept of feminist technology. The very idea of feminist technology at all is a result of the intellectual and academic critique of technology. That critique had at its core the analysis and evaluation of technology in terms of the social. So it is understandable that feminists engaged with these critiques would develop their own critique, one that would help make technologies that would better women's lives, and that engineers and designers attentive to the social component in their work would want to create technology responsive to women's needs.

The process of developing technology for women's bodies and lives can take place without any insight or examination. But eventually, failed or inadequate or less than successful technologies and their effect on women's lives demand a critical assessment of how to create a feminist technology. To do this, it is necessary to give "feminism" a meaning that provides a general goal and is widely acceptable without being doctrinaire and inflexible. There are certain concepts applied to women as a social group that have to be included: their needs and capacities, what benefits them, effective empowerment, and gender equity. These concepts are basic without being essentialist and provide the touchstones for conducting and evaluating research. There is a caveat, however. Because gender is a cultural construct and performative, when scholars and activists set a feminist goal, they are both constrained and naive. They are constrained by their own construction and performance of gender and naive about what other future forms of construction and performance are possible (Butler 1990). This does not mean they are paralyzed. It does mean that concepts alone cannot make change. They must also look to spontaneous innovations in social behavior, sometimes very small, as additional guides to finding the directions for a future successful feminist technology.

The issues connected to technology are no less plentiful and controversial than those connected to feminism. Historically, even the definition of technology had to be reevaluated by feminists in order to juxtapose it with women. For example, machines not in the marketplace but in women's domains such as the household were linguistically and conceptually not technology. They were called "appliances" and not thought of as the same as machines in the marketplace even though they had the same components and required the same skills for operation. The reasons for and the ramifications of such naming are part of an entanglement of the social and the technological. The concept of the sociotechnical system allows the authors to examine the materiality of technology while being alert to the social matrix in which it is embedded. But there is no stasis here. Technology is continuously tested both by manufacturers and users in the marketplace, leading to continuous processes of adjustment. Gender social relations are continuously changing, and even biological gender, once seemingly immutable, is not stable. These shifts in both technology and gender make for relentless flux within sociotechnical systems.

Recognizing these instabilities, both Johnson and Layne provide generalized rubrics that address them. Johnson, in elucidating feminism as the social relations under scrutiny, starts with the idea of the improvement of women's lives. There are different kinds of improvement. Sometimes what is good for women is feminist. In other circumstances, it is what achieves gender equity. In other conditions, what favors them is the necessary criterion. Layne, in considering candidates for feminist technologies, suggests that they might best be differentiated on the basis of how much they improve women's lives. She suggests evaluating technology based on snapshots of continually changing sociotechnical systems. In any given moment in the chain of systems, then, some technologies can be said to improve women's lives a little, while others improve them a great deal. All such technologies can be said to be feminist in their own way.

What we know then is that there is not a single kind of entity that can be called a feminist technology. However, in the contemporary Western industrial world, there is certainly an ethos that permits and encourages social and political activists as well as engineers and designers to provide increasingly appropriate technology for improving women's lives based on the belief that women are still in a subaltern position. How women came to hold this position in the Western industrial world has been modeled by Ruth Schwartz Cowan in *More Work for Mother* (1983). In Western preindustrial nonelite agrarian society, both men and women worked in the household. They also both worked outside the household; that is, they had work tasks that connected them to the local economy. For example, men were predominant in agriculture; women were predominant in cloth-making. In both these arenas, men and women understood each other's tasks and tools and, for many reasons, such as illness or absence, were knowledgeable about the other's work activities. For the vast majority of people, preindustrial work life was less gendered than social and political life. It was the Industrial Revolution and capitalism that transformed work life and with it the gender roles of work life. For one thing, more men's than women's household tasks were outsourced. This gave men the opportunity to work in the marketplace. Because women's tasks were not outsourced, they needed to remain in the household to sustain domestic life.[1] These two sites (the marketplace and the household) of modern life became more discreet, more differentiated, and more gendered than before.

Although women participated intermittently in industrial capitalist society, they were not the people forging the new economy, which was the dominant force shaping the new society. Nevertheless, the nineteenth century saw improvements for women. Laws were enacted that bettered women's position in regard to property rights and suffrage. There were social movements that argued for perfect equality among men and women (John Stuart Mill's famous essay "The Subjugation of Women" was published in 1869). Yet in the first half of the twentieth century, women in the Western industrial world were still in a substantially subaltern position. After World War II, women's lives took an unexpected lateral turn. There was a configuration of technology and social relations that confined women's lives while claiming to improve them. With wartime technology no longer needed, industrial capitalism looked to other markets for their products. They found one in the domestic domain, where they could sell household appliances and family automobiles. In June 1945, with the war coming to a close, *Modern Plastics* magazine looked to the postwar marketplace. "[The modern housewife] anticipates that all the developments which create fighting weapons today will tomorrow furnish the power to drive her household forces" (Henthorn 2000, 176). The locus of postwar economic expansion moved from the battlefield to the home and from the military consumer to the household consumer. At the same time, there was a postwar elevation of hearth and home, so it was not unexpected that women, who since postagrarian times had run households, would be expected to be there and, in many cases, to return there. Yet the social gains of the previous women's rights movements were not lost. There was an implicit cultural ethos that women's lives should be made better. The manufacturers of new household appliances advertised that their appliances made women's lives better by creating more hygienic conditions (protecting the lives of children particularly [Sivulka 2001]), eliminating the grueling aspect of household labor, and affording women more leisure time (Henthorn 2000).[2]

But this apparent benevolence was not sufficient. Women did not consider their lives in the domestic domain good or even good enough. In her popular book *The Feminine Mystique* (1963), Betty Friedan called it the problem with no name. What had happened was that women began to see that the marketplace, not the household, was the engine of modern society. They wanted to design and run that engine along with men. They also wanted working lives that satisfied their abilities and talents, not merely intermittent jobs, catch-as-catch-can jobs, or helping-out jobs. Since the 1960s, the ideas of feminism have been incorporated in one version or another and in one way or another into the mainstream of Western, if not global, culture. There is an awareness of women that is different from what it was before the 1960s. There is a feeling that women are not satisfied with their lives and want to continue to improve them in new ways. There is a sense that women are being ignored, slighted, wronged, and subject to injustice.

In response, individuals and institutions look for remedies. The marketplace is as much touched by these notions as any other part of society. It is no surprise, then, that industrial capitalism enters the gender fray with technological remedies. But industrial capitalism has its own goals that can and have trumped whatever sense of responsibility it apportions for women's interests. It is feminist sensibilities among users and potential users that can change the market so that industrial capitalist interests become increasingly congruent with women's interests. In other words, if women, goaded or influenced by feminist activism, ask for technology that will better their lives—that is, become a market for such technologies—industrial capitalism will find it in its interests to develop these technologies.

Technologies for this market are made by engineers and designers, who live in both the technical and greater quotidian worlds. When they conceive of a technology, they are aware, no matter how imperfectly, that it will exist in the world. In other words, they have a vision not only of how the object will work, but also how it will work and be used in the world. They inscribe these visions into the object itself. Every technology has embedded in its technical content a script based on these visions (Akrich 1992, 205–24). The concept of scripting allows us to see that technological objects can be political, as Langdon Winner (1986, 19–39) has told us in another context. This means that feminist engineers and designers, whether men or women, will write a different script for a technology than, say, those who are not thinking about women at all, let alone how women's lives can be improved. The designers of the fighter plane cockpit could ask whether this cockpit fits most men or they could ask whether this cockpit fits most people. On the other hand, users of objects can edit or even rewrite the script. Women can take objects made for one purpose and use them for their own purposes. The telephone was scripted as a business device to convey short messages. Women used it as a means of breaking out of their nuclear household isolation and putting themselves into a community that could provide support and aid in crises and in everyday life.

The dynamic relationship between the social and technical cannot be characterized by causality or determinism. Rather, there are continuous alterations and adjustments of the components of sociotechnical systems that move the systems over time into what can be described as different or successive states. Because a system has so many parts or components, not always completely known, and the parts or components may have many kinds of interactions, it is not possible to predict the chain of states of a system. In other words, systems are contingent—that is, their metamorphosis is not bound by necessity; they are open to the effect of chance and free will. Although new components to a system can have substantial effect, it is not unusual for alterations in already existing components to lead ultimately to substantial change in sociotechnical systems and society.

Action that brings about change for women's lives has many sites and many forms in the sociotechnical system. An action, whether technological or social, is the introduction, alteration, tinkering with, or tweaking of components and/ or their interaction with other components of the system. An action need not be the creation of an original technology; indeed, it is just as likely to be the redesign of a technology (a useful discussion of redesign is in Latour 2008). Feminist action need not be the adoption of a fully formed coherent consensual feminist ideology. It can be subscribing to a principle, such as women should be hired in any job they can perform. Smaller social adjustments, such as variation in use, can constitute feminist action. Even unintended consequences of design or behavior, if co-opted, can be feminist action. Feminist action, then, is any action that disrupts and directs sociotechnical systems, leading to new configurations that further benefit women.

The long-term menstrual-suppressing birth control pill story shows how a technological object, the original birth control pill, can be altered not by any modification of its physicality, but simply by its use. That altered use is an action that benefits at least some women. The original birth control pill prevented pregnancy by suppressing ovulation; that is, the hormones in the pill kept a woman's ovaries from releasing eggs, which suppressed menstruation. Users discovered that taking the pill continuously through the month and not taking the placebo pills for the allotted seven days of the month prevented menstruation completely, which they considered a benefit. Their idiosyncratic use of the pill in effect remade a technological object. That remade technology eventually was packaged, emphasizing the derived use, and became the long-term menstrual-suppressing pill that was also a birth control pill. Several ideas about gender had developed in the decades between the two products that made the technology more marketable than before. One was that a lesser degree of sexual dimorphism was acceptable; women could be more like men and men could be more like women. Clothes, hairstyles, and adornments were becoming more and more unisexual. In other words, there were fewer cultural markers for each gender. The diminution of cultural markers makes it easier for some to pay less attention to biological markers. Yet for others it makes biological markers even more important. The derived use of the birth control pill to suppress menstruation led to a new social component: women needed to have an opinion about whether they did or did not want to menstruate. Before the pill existed, no such opinion was needed. In other words, the use of the technology led to an evaluation of social attitudes involving biological and cultural sexual dimorphism, adding a new consideration to the already ongoing dialogue about gender and gender-bending.

While some actions improve women's lives, other actions also have the potential for worsening their lives. The introduction of the pregnancy test is an example of major action taken on behalf of women that did not better their lives as expected. The test is a foundational technology and exclusively for women. On these grounds alone its introduction would seem to be an action with good potential for improving women's lives. In addition, it is based on a modern, reasonably successful cultural premise: that science and technology are a better means of addressing and solving problems that had been previously addressed and solved observationally and intuitively. As a modern technological device, the pregnancy test brought with it all the expectations surrounding such devices: that it would be reliable and provide unequivocally the information it promised. Women users did not accept its reliability, which made it effectively unreliable, and it did not provide all the information women needed; for example, the possibility of and reasons for false positives. The test disrupted the previous system. Early private knowledge of pregnancy became an unstable system component because its authority is ultimately preempted by the medical establishment, whether the pregnancy is continued or terminated. In addition, the information in the system is inadequate, misleading the test takers as to the range of interpretations of the results and thereby fostering misunderstandings. In this case, women's lives are lessened to the degree that their expectations are unmet and their knowledge underfed.

The introduction of the breast pump is an action like the introduction of the pregnancy test: a foundational technology enters into an already existing system. It is different from the pregnancy test because in the case described, it is for women who work in the industrial marketplace and brings with it the idea of the public. The configuration (both social and spatial) of the marketplace precludes nursing mothers from caring for their infants there. This means that women need to care for their infants from a distance. The breast pump is one of the many things needed to provide that care. The marketplace breast pump brings with it important elements of public circumstances: spaces, business relations, and legal obligations. These are the obvious loci of action. But there are older associations that remain in the new configuration of circumstances around the breast pump, namely the social attitudes about breast-feeding, that place it firmly in the private and deny it a place in the public domain. Yet action by employers (J. P. Morgan Chase, CIGNA, and Kodak, to name a few)[3] that set up lactation rooms with breast pumps were not met by protests by those who held the older attitudes. These companies did not address their concerns before setting up the rooms. Every component of the sociotechnical system, even relevant components, need not be acted upon to achieve a goal. Some actions, sometimes predictably but sometimes unpredictably, have multiple effects. In this case, institutional action not only set up the lactation room, but also trumped opposition to the room before it came to the fore.

Re-scripting is an action that takes advantage of already existing technologies. Re-scripting is the additional vision by the designer or user that alters the technical or social aspects of the original script embedded in the technology. The case of Natracare is about re-scripting a technology already of benefit to women to make it a better technology for women. The original tampon holds a potential danger for women. It is made of standard bleached cotton, which may subject a woman who wears it to allergic and toxic reactions. By substituting organic cotton for bleached cotton, the re-scripted tampon eliminates that risk. In a different way, women's lives are expected to be improved by the re-scripted vibrating tampon whose motor acts to ease women's menstrual pain. In a further re-scripting by users, it functions as a sexual vibrator.

Rapex is another example of re-scripting tampons, but is different from the others in that it involves not only the woman who buys it and wears it, but also another component, the rapist. The intent of the designer is to empower women against rape. Yet the designer has not taken into account the dynamics of component interaction. Both forcible rape and murder are violent crimes (FBI 2007). The perpetrator of one should be considered more likely to commit the second crime than, say, noncriminals or nonviolent criminals. In other words, a rapist injured by Rapex is likely or more likely than others to react by murdering or further assaulting the rape victim. The action embedded in the technology is flawed in that it addresses one component without taking into account other crucial components of the system. This may be a defect of re-scripting. In re-scripting, elements of the original design linger on uncritically in subsequent iterations. A more successful technology would likely be derived by addressing rape as the initial design problem, allowing the designer to consider de novo the relevant components of the system in which rape is embedded.

It would appear from the discussion so far that action on the overwhelmingly complicated and volatile sociotechnical system has the potential danger of producing unwanted effects. On the other hand, informed and directed action decreases that danger and increases the chances of producing wanted effects and, thereby, a wanted change in a system. Such action made for effective change in the cases of long-acting contraceptives and the microbicides offering protection against STDs. The developers of these technologies are technoscientists (physicians, pharmacists, chemists, bioengineers, etc.). They are interested in a relatively safe technology that solves the medical problem. Also involved are policy makers who are potential purveyors (a kind of user) of these technologies on behalf of the state. Because these groups have their own interests, they are not the best actors for women, who are the ultimate users of these technologies. Women's interest requires that women have both optimum and maximum choices as persons[4] over their fertility and sexuality and that their choices result in minimum health risks (Hardon, this volume).

In order to assert their interests in the most unadulterated way, women users must have their own actor group. But this optimal solution is not available. Women users are too geographically dispersed and not in places that would facilitate the creation of an advocacy group. The actor group needed is one that has expertise not only among technoscientists, but also among policy makers. In addition, such a group would have to align itself with women users' needs, desires, and rights. It is understandable, then, that women's health advocacy groups consider themselves surrogates for women users, and women users can and have acknowledged the role of women's health advocacy groups by publicly supporting particular campaigns.[5]

In the case of the Norplant implant, action was undertaken by members of women's health organizations who had determined the dangers to women's health and well-being caused by the technology. By gaining the cooperation of the pharmaceutical industry, they were able to alter the procedures around the technology. In other words, an informed and directed re-scripting of the already existing technology was an action that made a formerly unacceptable technology acceptable. In this case, re-scripting required a large-scale organized effort on the part of women guided by specific and appropriate principles.

Influencing the script for a technology is more likely when the technology is in the process of being designed and when the diverse actor groups involved actually coalesce (as in the antifertility vaccine and the microbicides cases). There are often obstacles to such coalescence. There can be disagreement among groups as to who actually represents women users. This can remain an unresolved and underlying issue in ensuing dialogue and in the ultimate negotiation of direction and change. Certainly there are instances when women users' goals are in direct conflict with other groups' goals. Government family planning workers want a technology that will limit population growth; users want to choose among several technologies that will offer them options for managing their fertility and sexuality without risk. Technoscientists in their clinical testing are more concerned with populations; it is in women users' interests to focus on individuals in test results. Even advocates' interests cannot be entirely congruent with users. Advocates can be expected to have their own agenda. For example, they must maintain their long-term relationship with other actor groups. Also, they may have a cultural and social sensibility different from users; that is, they may be from industrial societies whereas users are from traditional societies, and they may be from moneyed classes whereas users may be from classes that are in poverty.

But these actor groups are not inflexibly bounded; their boundaries shift. Interest positions are subject to change. The membership of groups overlap; family planners or technoscientists can also be sympathetic to the idea of women's right to choice and well-being. When these actor groups are engaged in dialogue, the dynamic underlying shift and overlap enable collaboration. The groups are able to negotiate to bring the users' sense (knowledge, needs, preferences, views, and experiences) of the technology and its social context to production, testing, and marketing. When aspects of sociotechnical systems are configured through negotiation, original technologies as well as iterations of earlier technologies are more likely to be successful because they are embedded in a more appropriately complex context.

The action of negotiation is characterized by conciliation among diverse professionals. But what if a feminist agenda directed the process from the outset? An initial feminist agenda in the development of a technology requires more than that actor groups be advocates of or sympathetic to women's social and political rights. It demands that the engineering or design project solves a feminist problem from the outset. This means addressing an aspect of how gender relations can be improved in society and culture. Bronet and Layne (this volume), in describing this kind of action, compare it to green design or design for the disabled. In other words, instead of addressing industrial market goals first and those of a disadvantaged group second, the goals are addressed simultaneously. In feminist-agenda action, engineers, designers, manufacturers, and marketers of technology attempt to mitigate an aspect of gender inequity in the very places of inequity. For example, in the wage workplace, the project might be producing a technology that would ensure equal pay for equal work. In the political arena, it might be designing a technology that would enable women to vote in numbers equal to men. In the household, it might be designing a technology that would reduce women's vulnerability to domestic violence.

Feminist agenda action has the advantage of being able to create technologies that do not yet exist in the planning or prototype or product stage. It recognizes or senses and addresses some injustice or rather the pain of the injustice, to follow Elaine Scarry's thesis on the body in pain (1985) briefly here. Writing about the role of sentience in the creation of objects, she says a chair is "not in the shape of the skeleton . . . not even the shape of pain-perceived, but the shape of perceived-pain-wished-gone. The chair is therefore the materialized structure of a perception; it is sentient awareness materialized into free-standing design" (1985, 200). By asking root questions, feminist-agenda action addresses the pain (to continue with Scarry's concept) of gender inequity directly in the places that it exists. Translating the questions into engineering or design problems enables in the most parsimonious way the creation of a particularly efficient and appropriate technology. In other words, although the marketplace economy is taken into account, it is not the overwhelming consideration in directing the creation of the technology. Such a technology does not have the disadvantages that often accompany a re-scripted technology, in which lingering features and embedded ideas of the original design remain to potentially plague the revised technology.

No matter how they are created, technologies as material objects are themselves forces of action because they carry ideas. Material objects convey cultural themes that are apprehended sentiently by both the maker of the object and the user. Although most cultural themes (for example, individualism among Americans, shame among the Japanese) are normative—that is, they are generally held—some themes are not commonly or widely accepted. These controversial themes are nicely introduced in objects because their very objectification allows them to avoid being discussed, debated, or argued, which can be potentially abrasive interactions of social relations. In other words, controversial themes can be accepted or rejected in the object without risk to other social relations.

The evaluation of an object's message can be seen by imaginatively expanding on the object/space story of the diaper-changing table in the men's bathroom. The man seeing the table begins an internal dialogue. "This must be a women's bathroom because only women change babies' diapers." "But, no, I see now it is in the men's bathroom." "This must mean that someone out there thinks that men change babies' diapers." "This must not be an idiosyncratic thought because the bathroom is in a public facility." "A lot of people must think men change babies' diapers." "It must be that men change babies' diapers." The mere existence of the object/space carries the new gender relations message. Institutional power is an effective way of launching the message simply by placing it into use. In the case of the breast pump/lactation room, once J. P. Morgan Chase established the facility, it was accepted even by those who held that breast-feeding did not belong in the marketplace. By fiat, breast-feeding was in the marketplace.

What do we know now about feminist technologies? We know a few simple things. Feminist technology is not merely about technology for women's bodies. Nor is it technology originally and perhaps ingenuously designed for men and adapted for women. And it isn't technology gussied up to reflect a male or even a female sensibility of femininity. We also now know a few complicated things. Feminist technology is an instrument for changing current unsatisfactory gender relations and societal norms so that women have better lives. But feminist technology is part of a roiling entity. Modeled as a sociotechnical system, that entity can be understood as infinitely complex, consisting of components that are in flux and interconnected in unstable ways. Action that alters the components and the interactions of the components can produce substantial change in the system.

What we don't fully know is which actions are most likely to be successful in improving women's lives. Sometimes it is the creation of a technology; sometimes it is in the adjustment of a technology; sometimes it is in the social relations connected to the technology (for example, in use). The scale of action does not always predict the scale of the results. A big action might make for little change, while a little action might make for big change. But what seems to be fundamental is that action informed one way or another by feminist ideas is most likely to produce sociotechnical systems in which women's lives are improved.

NOTES

I thank Deborah G. Johnson for continuing critique and Linda Layne and Ellen Esrock for insightful comments.

1. The responsibility for raising children fell to women not only because of their unique connection to children, but also because they were tied to home tasks more than men were. In agrarian society, children, particularly boys, were cared for by their fathers even in preteen years as they assisted them in male chores.

2. Ironically, women's labor in the household did not decrease; it merely changed. Women, with their appliances in hand, now did chores previously done by servants, and added chores, such as chauffeuring, that had not even existed earlier (Cowan 1983).

3. Thirteen states and the District of Columbia have laws related to breast pumps and lactation rooms in the workplace (NCSL 2008).

4. Their role as persons is different from their role as citizens. In other words, individual well-being and rights are considered to take precedence over societal responsibility.

5. Sometimes women's health advocacy groups may not represent women users. One example is the case of the breast-feeding advocacy group that opposed bottle-feeding in certain traditional societies, disregarding information about the reasons the women in those societies regarded their choice of bottle-feeding as rational.

References

Akrich, Madeleine. 1992. "De-Scription of Technological Objects." In Wiebe Bijker and John Law, eds., *Shaping Technology/Shaping Society.* 205–24. Cambridge Mass.: MIT Press.

Butler, Judith. 1990. *Gender Trouble.* New York and London: Routledge.

Cowan, Ruth Schwartz. 1983. *More Work for Mother: The Ironies of Household Technology from the Open Hearth to the Microwave.* New York: Basic Books.

Federal Bureau of Investigation (FBI). 2007. http://www.fbi.gov/ucr/cius2007/offenses/violent_crime.

Friedan, Betty. 1963. *The Feminine Mystique.* New York: Dell.

Henthorn, Cynthia. 2000. "The Emblematic Kitchen: Labor-Saving Technology as National Propaganda, the United States, 1939–1959." In S. Gorenstein, ed., *Research in Science and Technology Studies: Gender and Work,* 153–88. Stamford, Conn.: JAI Press.

Latour, Bruno. 2008. A Cautious Prometheus? A Few Steps toward a Philosophy of Design (with Special Attention to Peter Sloterdikijk). Keynote lecture for the Networks of Design Meeting of the Design History Society. Falmouth, Cornwall.

National Conference of State Legislatures (NCSL). 2008. http://www.ncsl.org/programs/health/breast50.htm.

Scarry, Elaine. 1985. *The Body in Pain: The Making and Unmaking of the World.* New York: Oxford University Press.

Sivulka, Juliann. 2001. *Stronger Than Dirt: A Cultural History of Advertising Personal Hygiene in America, 1875–1940.* New York: Humanity Books.

Winner, Langdon. 1986. *The Whale and the Reactor.* Chicago: University of Chicago Press.

Contributors

JENNIFER AENGST is a graduate student in the Anthropology Department at the University of California, Davis. Her research investigates the interactions of family planning programs and the Tibetan medical system in Ladakh, India, focusing specifically on issues of sexuality, risk, and morality. She has a master's of international development and human rights from the University of Denver and a bachelor's in philosophy from Colby College.

MAIA BOSWELL-PENC is an assistant professor of women's studies at the State University of New York in Albany, where she has been teaching since 1999. Her book *Tainted Milk: Breastmilk, Feminisms, and the Politics of Environmental Degradation (2006)* builds on her interests in environmental justice, gender, and the politics of justice. Other teaching and research interests include disability studies and feminist pedagogy.

KATE BOYER received her Ph.D. in cultural and feminist geography from McGill University in 2001. She taught in the Department of Science and Technology Studies at Rensselaer Polytechnic Institute between 2002 and 2007, and is now a lecturer in the School of Geography at the University of Southampton. Her interests include questions of gender and technology in the workplace, the cultural politics of design, and the processes through which space and place come to have meaning. Her work has appeared in *Gender, Place and Culture; Antipode; Space and Polity; Social and Cultural Geography;* and *Urban Geography.*

FRANCES BRONET is dean of the School of Architecture and Allied Arts at the University of Oregon. She specializes in multidisciplinary design between architecture, engineering, science studies, dance, and electronic arts. She completed a series of funded interactive full-scale architecture, construction, and dance installations with the Ellen Sinopoli Dance Company. Recent publications include "Beating a Path: Design and Movement" in *Performing Nature: Explorations in Ecology and the Arts;* "Product Design and Innovation:

The Evolution of an Interdisciplinary Design Curriculum" in the *International Journal for Engineering Education;* "Design in Movement: The Prospect of Interdisciplinary Design" in the *Journal for Architectural Education;* the 2001 Contemporary Justice Review on Design and Justice; and "Quilting Space: Alternative Models for Architectural and Construction Practice" in *Knowledge and Society* 12 (2000) on a liberative teaching model for studio learning based on the experience of marginalized students. She is the recipient of numerous teaching awards, including the 2001 Carnegie Foundation for the Advancement of Teaching New York Professor of the Year.

SHIRLEY GORENSTEIN is the author or editor of twelve books in the fields of science and technology studies and American archaeology. She edited and wrote the theoretical introductions to the *Research in Science and Technology Studies* volumes: *Material Culture* (1996), *Knowledge Systems* (1998), and *Gender and Technology* (2000) (coauthored introduction). Also, she coedited and wrote the theoretical introduction to *Greater Mesoamerica* (2000). She was the founding chair of the Department of Science and Technology Studies at Rensselaer Polytechnic Institute, where she served for twelve years as chair and eight years as associate dean for graduate studies in the School of Humanities and Social Science.

ANITA HARDON is professor of health and social care at the Faculty of Social Sciences of the University of Amsterdam. She graduated cum laude in medical biology from the University of Amsterdam (1984), specializing in tropical hygiene, science dynamics, and medical anthropology and completed a Ph.D. in the Department of Cultural Anthropology/Non-Western Sociology at the University of Amsterdam studying "Medicines, Self-Care and the Poor in Manila" in 1990. She subsequently coauthored *Drug Policy in Developing Countries* and formulated the Community Drug Use Project (1992–94), which resulted in a series of studies in Pakistan, the Philippines, Mali, and Uganda. More recently, she has undertaken a number of case studies on women's subjective experiences of reproductive health. She has been adviser on user issues to agencies involved in the development of reproductive technologies (such as the World Health Organization and the Population Council) and led an eight-country study on reproductive rights in practice, assessing the way in which family planning services worldwide are providing contraceptives to women and men.

DEBORAH G. JOHNSON is the Anne Shirley Carter Olsson Professor of Applied Ethics in the Department of Science, Technology, and Society in the School of Engineering and Applied Sciences of the University of Virginia. She received the John Barwise prize from the American Philosophical Association in 2004; the Sterling Olmsted Award from the Liberal Education Division of the American Society for Engineering Education in 2001; and the ACM SIGCAS Making a Difference Award in 2000. She is the author/editor of four books: *Computer Ethics* (Prentice Hall, first edition 1984, second edition 1994, third edition 2001); *Computers, Ethics, and Social Values* (coedited with Helen Nissenbaum, Prentice Hall, 1995); *Ethical Issues in Engineering* (Prentice Hall, 1991); and *Ethical Issues in the Use of Computers* (coedited with John Snapper, Wadsworth, 1985). She has published more than fifty papers in a variety of journals and edited volumes. She coedits the journal *Ethics and Information Technology* and coedits a book series on Women, Gender, and Technology for the University of Illinois Press.

LINDA L. LAYNE is the Hale Professor of Humanities and Social Sciences and professor of anthropology in the Department of Science and Technology Studies at Rensselaer Polytechnic Institute in Troy, New York. Her books include *Motherhood Lost: A Feminist Perspective on Pregnancy Loss* (Routledge, 2003), *Consuming Motherhood,* edited with Janelle Taylor and Danielle Wozniak (Rutgers University Press, 2004), and *Transformative Motherhood: On Giving and Getting in a Consumer Culture* (New York University Press, 1999). She is coproducer of a television series, *Motherhood Lost: Conversations*, with Heather Bailey at George Mason University Television. Her current research is on single mothers by choice.

DEANA MCDONAGH is an associate professor of industrial design in the School of Art and Design at the University of Illinois at Urbana-Champaign and is part of the faculty at the Beckman Institute of Advanced Science and Technology. Her interdisciplinary research concentrates on developing empathic design research approaches for designers to ensure more effective product outcomes. She focuses on user-product attachment, design and emotion, and developing appropriate technologies for females, aging users, and users with disabilities. She coedits the *Design Journal* and is the guest editor for the special edition (Design and Emotion) for the *Journal of Engineering Design*.

SHARRA L. VOSTRAL is an associate professor, holding a joint appointment in the Program of Gender and Women's Studies and the Department of History at the University of Illinois at Urbana-Champaign. Her research centers on the history of technology in relation to gender and women's bodies and the ways in which material artifacts function in people's everyday lives. Her book, *Under Wraps: A History of Menstrual Hygiene Technology* (Rowman and Littlefield, 2008), examines the social and technological history of sanitary napkins and tampons through the lens of passing and the effects of technology on women's experiences of menstruation. Her current research explores toxic shock syndrome and its relationship to tampon use during the early 1980s.

Index

abortion, 15, 91, 94, 97, 98, 105, 108, 109, 192, 208; clinics, 91; 1st trimester, 102; induced, 101; legal, 114n29; rights, 155; spontaneous, 8, 101, 102, 104, 108. *See also* pregnancy loss

absenteeism, 61, 83n41, 131

abuse, 106, 161, 164–67; abuser, 7; of reproductive rights, 73; sexual, 109; by state, 169

Academy for Entrepreneurial Leadership, 198

access, x, 11, 20, 77, 105, 110n1, 115n40, 126, 129, 130, 159, 161, 180, 200

accessibility, 11–12, 22

accuracy, 89, 94, 100, 99, 111n5, 113n24–25. *See also* efficacy

Actor-Network Theory (ANT), 39, 210–11

advertising, 99–100, 104, 120–21, 124–25, 128–29, 136, 138–39, 142, 144, 151, 157, 181, 185, 205

Aengst, Jennifer, 18, 55–88, 104, 179

aesthetics, 146, 180, 191, 201; women's, 4

Africa, 70, 155

African American Feminism, 72–73

African American women, 24, 72–73, 83n41, 122, 123, 184

age, 22, 60–61, 81n22, 122; child-bearing, 75, 93, 102. *See also* teens

aircraft, 187–88, 206

Akhter, Farida, 160

Akrich, Madeleine, 136, 154, 206

Allerton, Haidee, 130

American Society for Engineering Education, 188, 194n8

American Society of Civil Engineers, 188

American Society of Engineering Educators, 194

American with Disabilities Act, 11

Andrist, Linda C., 61

Angell, Marcia, 75

Anita Gorg Institute for Women and Technology, 189

anonymity, 15, 20–21, 98. *See also* privacy

anticonsumerism, 127

antifeminism, 5, 7, 18, 43, 44, 50, 52n2, 70, 71, 76, 143, 154, 161, 166, 169, 179

antifertility vaccines, 23, 79n10, 154, 161, 173; call for ban of, 167; injectable, 169; intrinsic characteristics, 165; long-acting, 169; oral, 169; short-acting, 169

antitechnology, 127

Apple, Rima, 121

appliances, 37, 38, 138, 204, 205, 213n2

appropriated technologies, 64–66, 82n34, 92, 103, 104, 151, 152, 206, 207

appropriate technology, 42, 78, 107, 110, 199, 204, 212

architecture, 15, 180, 199; green, 14, 191

Argentina, 89, 101, 111n9

Ark, community charter school (The), 183

Armbruster Bell, Wendy, 8

artifacts, 6, 25n5, 36–54, 119, 120, 143, 151, 161, 181, 185, 191; as agents of change, 125, 193; feminist, 47–49; gender of, 154, 193; politics of, ix, 179, 180, 195n11; sexism inscribed on, 182

Association of Collegiate Schools of Architecture, 194

athletes, female, 61, 66

Australia, 81n29, 155, 162

authoritative knowledge, 94–97, 108, 208

automobiles, 191, 194n3, 205; as equipment designed for men, 191; redesign for women, 183

autonomy, 13, 104, 107, 125, 174. *See also* freedom; independence

Avent, Phillips, 124, 126

babies, 1, 17, 81n30, 111n9, 120, 121, 195n10, 212; feeding, 8–9, 18, 19; transporting, 10–12. *See also* breast-feeding

baby bottles, 8, 15, 17, 18–19, 39, 48, 82n40, 119, 120, 124, 209, 214n5

Baby Buddha, 9

baby-changing, 182, 195n10, 212. *See also* diapers

Backlash, 23

Bailey, Richard, 8, 139, 140

Baker L. D., 97, 100, 102, 105, 111n5, 114n32

Bali, 66

Balka, Ellen, 50

Bangladesh, 156–59

banishment, 15–16, 129, 132

Barbiere, Stacy, 149

Barnes, Shellie, 72

Barzelatto, J., 162

bathrooms, 92, 127, 129, 132n12, 144. *See also* toilets

Bazelon, Emily, 124, 131

Beastie Boys, 150

Beattie, K., 159

Belkin, Lisa, 16, 19

Bell, Genevieve, 5, 19

Bellamy, Robert V. Jr., 7

Benagiano, G., 167

benefit, 22, 23, 29n33, 56, 63, 75, 76, 79n11, 91, 114n32, 181; of breast pumping, 131, 138, 144, 174, 187; health, 55, 56, 167; of home pregnancy test, 94, 100, 104–6; of nursing, 122, 123; sexual, 149; of technology, 3, 89, 124, 179; to users, 166; vs. choice, 70; to women, xin3, 4, 6, 10–16, 100, 108, 110, 124, 151, 171, 180, 203, 207, 209

Berer, M., 155, 167

Betty Friedan, 206

Bezanson, K. A., 168

Bijker, Wiebe E., 38, 39, 48

bioengineering, 121, 167, 210

biological differences, 71, 128, 132

biomedical, design, 189; knowledge, 97

birth control methods, 23–24, 28n19, 70, 75, 105, 109, 155; cervical rings, 79n6; Dalkon Shield, 156; Depo Provera, 56, 73, 79n11, 83n41, 156; development process, 164; diaphragm, 75, 143; efficacy, 110, 126, 162, 165–66, 170–72; emergency contraception, 109; long-acting, 154–56, 160, 210; for men, 69–70, 77, 83n45; men and women responsible for, 28n26; menstrual suppressing, 55–88; method failure, 158; oral vaccines, 168; pills, 77, 78, 95, 207; provider-controlled, 168; safety, 56–58, 110, 156; spermicidal, 171; subcutaneous devices, 119; user-controlled, 160; women-controlled, 110, 168

BlackBerry, 131

blighted ovum, 97

Blum, Linda, 121, 127, 132n5

Bobel, Christina, 127, 133n11

bodies, female, 90, 96, 119, 203, 213; difference in size, 137, 140, 183, 199; male, 199; touching, 140, 142

bodily fluids, 27n16, 61, 62, 81n25, 90, 126–28

Bongaarts, J., 160

Boston Women's Health Book Collective (BWHBC), 91, 93, 94, 107, 114n34, 119, 122, 124, 132n4

Boswell-Penc, Maia, 24, 119–35, 183

Bott, Tim, 16

Boyer, Kate, x, 17, 24, 119–35, 183

brainstorming, 17, 200

Brasner, Shari, 61

Braunstein, Grenn, 90

Bray, Francisca, 120

Brazil, 56, 81n30, 157, 158, 162, 166

breakthrough bleeding, 58

breast feeding, 18, 19, 29n34, 63, 120, 122, 125, 127, 129, 132, 132n5, 191, 209; advocacy, 127, 214n5; breast-feeding-friendly policies, 131; laws protecting, 126; rates of, 121–24; as scandalous, 126; social attitudes about, 209; socially unacceptable, 126

breast milk: benefits of, 121; contaminated, 128; cost of, 121; expressed by hand, 122, 132n4; fear of, 128; nutritional value, 121

breast pumps, 8–9, 18, 77, 123, 154, 183, 208, 213, 213n3; advertisements, 120, 124, 125, 128; cost of, 122, 130; discreet, 124; ease of use, 124; "hands-free" model, 124; lease of, 122; manufacturers, 120; Medela, 24, 124; noise of, 126, 128; politics of design, 126; quiet, 124; source of anxiety, 128; users, 120

Brillo Magazine, 150

British Institute of Medical Laboratory Sciences, 112n14

Brodie, Janet, 138

Brody, Jane E., 63

Bronet, Frances, 5, 16, 24, 126, 179–96, 201

Brown, Murphy, 95

built environment, 11–12, 15–17, 21, 26n11, 48, 49, 120, 127–28, 130, 132, 183, 189

BUKO Pharma-Kampagne, Germany, 165

Burgess, Gemma, 21

Burill, Ives Marie Paul Jean, 138

Bush, Corlann Gee, 1, 3, 13, 25n2

Business Week, 22

Butler, Judith, 46, 203

buying power, 22, 89, 199

caesarean sections, 63, 81n27–30

California, 67, 99; College of Arts and Crafts, 191; Department of Health, 109, 115n42; Maternal and Child Health Department, 114n32

Calkins, John, 149

Callon, Michel, 39, 40

Campbell, Kim, 21

Campbell, Nancy, xiin6, 197

Canada, 8, 26n11, 57, 61, 89, 91, 101, 115n41

cancer, 90, 111n4; and pregnancy, 91; risk of, 55, 58

Cantor, Nancy, 195n12

capitalism, 62, 70, 71, 206

care work, 120, 131

Casper, Monica, 109, 115n43

Catholicism, 60, 81n24

Catley-Carlson, M., 166

Center for Interdisciplinary Research on the Family at Cambridge, xin1

changes: cultural, 132; cyclic, 70; legislative, 132; material, 132; policy, 132

Charry, Tamar, 104

Chesler, Giovanna, 61, 70

chicken-egg problem, 21, 45–46, 49, 51–52

child care, 8–12, 15–16, 20, 21; on site, 131; women and men responsible for, 16. *See also* babies

Child Health USA, 122–23

Chile, 162

China, 89, 101, 111n9, 115n37, 115n40

choice, 77, 108, 119, 124, 126, 127, 132, 144, 172, 210; consumer, 59, 61, 64; freedom of, 167; importance of, 132; of infant nutrition, 119, 124; informed, 108, 169, 172; and liberal feminism, 63, 70, 108, 167; limitation of, 144; natural, 126; reproductive, 57, 59, 64, 94, 126, 155, 159, 165–67; sensible, 172; user, 165; vs. benefit, 70; vs. rights, 80n14

churches, 15–16, 60, 66

CIGNA Insurance Company, 209

Clarke, Adele, 90, 109, 115n43

Claro, A., 156

class, 22, 64, 70, 77, 93, 101, 120, 123, 130, 181, 211; low-income, 21, 77, 130; middle, 20, 78, 101; poor, 183; upper, 78; upper-middle, 64

Clayton, Johnathan, 151

Cleveland Clinic's Health Information Center, 64

126, 192–93; curricula, 2; democratic, 89, 201; for the disabled, 2, 189, 211; ecofeminist, 152, education, x, 2; for elderly, 189; for equality, 5; ergonomic, 5, 181, 183, 199; feminist, 7, 40; of feminist products, 24; of feminist technology, 7, 24, 26n10; field, 126; goal of, 2; green, 9, 29n33, 191, 192, 211; human-centered, 199; interdisciplinary, xi, 24, 179, 184; for men, 4, 5, 6, 7, 46; for military use, 5, 46, 187; need-based, ix, 10, 23; neuter, 137; nonfeminist, 141; optimal, 137; politics of, ix; process, 2, 6, 17, 24, 45, 92, 120, 126, 170; production of, 66; programs, 87; of reproductive technologies, 156; sexist, 11, 12; social engagement of, 194; social scientists' role in, 171; of space and places, xin3; students, 26n16, 179–202; studios, 180, 188; sustainable, 199, 201; teaching, 24, 179; technological, ix, 8; universal 137, 181; urban, x; user-centered design, 24, 131, 154, 164, 168, 170, 174, 199; user-controlled, 160; user-friendly, 9; user-oriented, 174; woman-friendly, 201; for women, 3, 4, 5, 6–13, 27n15; workplace, 120, 130. *See also* feminist design

Design, Culture and Society, x, xin6

designers, ix, xin3, 5, 12, 16, 17, 24, 154, 204, 206, 211; industrial, 10, 197, 199; intentions of, 21, 137; sex of, 7, 8

design studio, 180; goal of design studios, 188

deskilling, 89, 107, 108

Despois, Emilie, 17

developing world, 18–19, 77

diapers, 17, 182, 195n10, 198, 212

differences: among users, 92, 199; among women, 55, 58, 78, 130, 180

Dioxin, 144, 146

disabilities, 11–12, 79n9, 137, 183, 194n3; access to public spaces, 11; women with, 66, 194n1

Disability Rights Movement, comparison with, 2, 12

discrimination, 21, 127, 132n9, 133n15

disempowering, 89, 154, 159, 161, 173, 174. *See also* empowering

disgust, 62, 82n37, 126, 150, 181

disposable products, 138, 140, 146; alternatives to, 71–72, 106, 143–44; applicators, 140; landfills, 82n40

Diva-cup, 143–44

diversity, 24, 26n10, 169, 174; among actor groups, 210; among designers, 126; of feminisms 18, 69, 74, 77; among students, 197; among users, 92, 126, 156, 173; among women, 3, 18, 26n10, 97, 124, 152, 166, 173, 179

Domosh, Mona, 130

Douglas, Mary, 126, 128

Dreyfuss, Henry, 199

drugs, 60, 147; companies, 55, 58, 59, 64, 79n4, 160; "me-too," 66, 79n5; over-the-counter, 94, 112n16, 114n33; stores, 105, 112n18, 115n38, 113n19, 140

Duden, Barbara, 103, 114n30

e-bay, 122

education, for patients, 101; of designers, 24, 144, 179–96; women's access to, 180; of women users, 106 9, 146, 167

efficacy, 89, 97, 100, 110, 111n5, 126, 162, 165, 166, 170–72; long-term, 173

efficiency, 61, 71, 128

Eglash, Ron, 66, 82n34, 103, 180, 183

Egnell, Einar, 124

Egypt, 81n24, 90

Ehlers, Sonette, 150, 151

ejaculation, 69, 70

electronics, 7, 22, 28n20–24, 119, 131, 147, 200

Ellertson, Charlotte, 57, 58, 59, 60, 61, 66, 79n9, 80n12–13, 82n31, 104

embarrassment, 113n21, 128–29, 139, 150

empowering, 2, 3, 13–14, 50–66, 68, 72, 119,

137, 141, 142, 150, 151, 155, 161, 169, 171, 172, 179, 188, 203, 209; empowerment, 41. *See also* disempowering

endometrial ablation, 83n41

endometriosis, 56, 57, 75, 79n9, 149

engineering, 126, 180, 184, 187, 197, 212; biomedical, 189, 194n8; design for disabled, 189; design for elderly, 189; education, 188, 198; environmental, 188, 194n8; students, 195n9

Engineering Projects in Community Service (EPICS), 189

engineers, 24, 137, 142, 145, 146, 195n9, 199, 204, 206, 211

enhancement technologies, 61–62

entrepreneurs, x, 28, 28n25, 144, 198, 200; entrepreneurial, 201; women, x, 7–10

environment, 22, 68, 82n40, 162, 180, 184; movements, 2, 4, 14, 23, 143, 144; toxins, 102. *See also* green

e.p.t., 91, 94, 100, 104, 111n6, 113n24

equality, 5, 6, 18, 21, 41, 42, 43, 47, 48, 49, 52n4, 73, 137, 180, 181, 205, 211, 212. *See also* inequality

Equality and Human Rights Commission, 21

Equal Rights Amendment, 193

equity, 6, 41, 42, 43, 46, 47, 48, 49, 50, 70, 136, 137, 179, 180, 184, 185, 186, 188, 190, 192, 199, 200, 203, 204. *See also* inequity

ergonomic, 181, 183, 199

Erol, Maral, 80n18, 80n21, 82n31

essentialism, 36, 41, 67, 69, 74, 77, 82n39; essential, 142, 180; essentialist, 18, 71, 203

Esser, Judith 142

ethical decision making, 146, 152

ethnicity, 66, 83, 93, 120, 121, 199

Eubanks, Virginia, 150

eugenics, 73, 160

Europe, 91, 155

European Centre for Health Policy, 23

Evans, Jenni, 151

Evans-Smith, Heather Gail, 103

everyday life, ix, 25n3, 38, 62, 161, 207

expertise, 37, 38, 126, 197, 210; experts, 56–58, 94, 138, 184, 186, 190, 200

Extraordinary Women Engineers, 188–89, 193, 195n9

fabrication laboratory, 201

Facebook, 50

Faircloth, Charlotte, 18

false positives, 90, 102, 208

Faludi, Susan, 4, 22, 23, 27n13

family planning, 7, 59, 64, 91, 155, 158, 167, 169, 211; centers, 105, 109, 114n30; choice of, 165; coercive 156, 161; counseling, 105; men's responsibility for, 160; programs, 156–57, 160–61; providers as users, 167; women's responsibility for, 160. *See also* birth control methods; contraceptives

Farrell-Beck, Jane, 137

Federal Bureau of Investigation (FBI), 209

femininity, 26, 27n13, 44, 61, 62–3, 64, 67, 68, 73, 74, 149, 213; feminine, 4, 6, 37, 38, 64, 68, 140, 150; and technology, 3, 26, 42, 136

feminism, 203; African American, 72–73; backlash against, 2, 23; cultural, 71; Cyborg, 74–75; definitions of, 2, 3, 7, 14, 18; diversity of, 58, 78, 179, 180; essentialist, 180; existential, 73–74, 83n43; ecofeminism, 71–72, 143; first wave, 4, 13, 205; goals of, 203; liberal, 18, 70, 142, 167, 169, 180; radical, 161, 180; second-wave, 142, 155, 206, Socialist, x, 7–10, 28

feminist, agenda, 190, 211; attitudes toward home pregnancy test, 94; design, 24, 190, 193; design critics, 193; designers, 75, 179; design principles, 143; design process, 7; movement, 3, 142, 199; pedagogic innovations, 186; washing, 3, 4, 57, 142, 151

Feminist Inventor's Studio, 24, 199–202

feminist technology: as artifacts, 47–49, 206; awards for, 24, 185–86; criteria for, 2, 11, 14, 36–54, 77, 106–9, 110, 181, 183, 204–13; definitions of, 1, 3, 6–7, 11, 14–15; designed by men, 7, 11–12; designed by sexists, 10–12; evaluating, 14, 204; foundation for the promotion of, 24; lack of consensus about, 14–15, 75–78; ranking of, 204; rescripting, 136–52; as sociotechnical systems, 59–51, 119–34, 211–13; teaching, 179–202; typology of, 14; used by men, 11

Feminist Technology Assessment (FTA), 14, 23, 78, 109, 110

Fertility Plus Organization, 91

fetal personhood, 103, 114n30

field studies of technology use, 158, 161

Fink, Jennifer L. W., 63

Finland, 158, 162

FINNRAGE (Feminist Network for International Resistance to the New Reproductive Technologies and Genetic Engineering), 161

Firestone, Shulamith, 83n43

Fisher, Jennifer K., 96

Foley, Bill, 187

formula, 18–19, 121. See also bottle-feeding

Fox, Mary Frank, xi

France, 21

Frederick, Christine Ann Nguyen, 48

free bleeding, 144

freedom, 4, 20, 27n16, 44, 57, 141; of choice, 167; physical, 136; reproductive, 74; spatial, 119, 124, 125, 131; temporal, 131. See also autonomy; independence

French Parity Law, 20

friends, 8–10, 71, 92, 99, 107, 114n31, 120

Frye, Marilyn, 41

functionality, 9, 42, 46, 47, 48, 49, 124, 130, 149, 152, 199, 201, 209

funding, 24, 109, 157, 167–69, 184, 187–88, 193, 194n7, 198, 201

Furst-Dilic, Ruza, 39

Gabelnick, H. L., 166

Gabriele, Gary A., ix, 180, 194n4

Garcia-Moreno, C., 156

Gawande, Atul, 79n2

Geisler, Cheryl, 6

Gelis, Jacques, 90

Gemzell, C. A., 90

gender, 21, 120, 194; bending, 208; bias, 179, 187, 194; changing, 204; discrimination, 21; equality, 5, 6, 21, 137, 205; equity, 6, 47, 70, 136, 179, 184, 185, 188, 192, 199, 200, 203; gap, 6, 20, 21; inequities, 201, 211, 212; neutral, 194n2; politics, 179, 180; roles, 185; and science, 180; system, 19; and technology, 120, 137, 154, 180

gender and technology, 120, 137, 154, 180; co-construction of, 38–40

gender and technology studies, x, 2, 197

gender and women's studies, 1, 3, 150, 197, 198, 200, 201

Gender Equality Duty, 21

gender scripts, 154, 165; antifeminist, 161; disempowering, 159; empowering, 161; feminist, 194n11; rescripting process, 136–52, 155

Georges, Eugenia, 82n38

Geraghty, Sheela, 122

Germany, 98, 114n30, 142

Gilbreth, Lillian, 138

Gladwell, Malcolm, 60

Gogoi, Pallavi, 22

Goodridge, Jenny, 5

Gorenstein, Shirley, xi, xin7, 4, 6, 12, 13, 18, 19, 120, 203–14

Gottlieb, Alma, 66

Greece, 82n38

green, architecture, 191; design, 191, 192, 211; ranking system, 14; washing, 4

Griffin, P. D., 163, 164, 165, 167, 168

guns, 25n2; stun guns, 5

Gupta, S. K., 162

Gurdon, John B., 90

inequalities, 6, 22, 24, 72, 83n43, 120, 185, 198. *See also* equality

inequities, 6, 19–21, 24, 41, 45, 46, 51, 52, 201, 211, 212. *See also* equity

infertility, 93, 107, 109, 110, 174

information technologies, 6, 89, 94–99, 103–4, 106–9, 126, 146

informed choice, 24, 57, 108, 137, 143, 152, 168, 172

informed consent, 158, 159, 161, 164

innovations, 2, 5, 6, 7, 8, 11, 12, 14, 17, 22, 23, 24, 26, 28n21, 198, 203; architectural, 15; cultural, x; ecofeminist, 143–47; feminist pedagogic, 186; radical, 14, 23, 147

inscription, 36, 154, 181, 195n11, 206; of sexism on an artifact, 182. *See also* scripts

Intel, 5

interdisciplinary, xi, 184, 195n12, 197; among K-12 children, 184; design, 179, 184; feminist scholarship, 110

International Committee for Contraception Research, 162

International Conference on People's Perspectives on Population, 159

International Contraception, Abortion, and Sterilization Campaign (ICASC), 155

International Development Research Centre (IDRC), 168

International Doula Conference, 8

International Women and Health Meeting, 155

Internet, 7, 98, 99; access, 20, 126; shopping, 126; women users, 126

Inter-Uterine Devices (IUDs), 119

interventions, 128, 120, 154–78, 186, 193, 210; feminist, ix, 23, 186, 201; timing of, 161, 170

introductory script, 159, 161, 167, 169, 172, 173

intuition, 114n34, 199, 208

inventors, 137; men, 138; women, 138, 142

investing, 20–21, 30n37, 102, 113, 143, 145, 146

invisibility, 138, 139, 144, 184, 185. *See also* anonymity; concealment; privacy

in vitro fertilization (IVF), 102, 109

irreversibility, 110, 164

Islam, 81nn24–25

iterative process, 17, 186, 209, 211

Ivory Coast, 70

Jamaica, 81n24

Jans, Lita, 194n1

Japan, 19, 20–21

Jayaraman, K. S., 168

Jenson v. Eveleth Taconite Co., 27n16

Johansson, E., 160

Johnsen, Michael, 93, 113n19

Johnson, A., 97, 111n5

Johnson, Deborah G., xi, 2, 4, 6, 18, 21, 36–54, 151, 204

Johnson & Johnson, 142

Johnston-Robledo, Ingrid, 57, 58, 80n15

Jones, W. R., 162

Jordan, Bridgette, 95, 96, 106

journalists, 92, 155

J. P. Morgan Chase, 130, 209, 213

Judaism, 16, 66, 81nn24–25, 186

Juno, 95, 115n35

Kammen, J. van, 154

Kathleen Landor-St. Gelais v. Albany International Corporation, 133n12

Kauffman Foundation, 198

Kaunitz, Andrew, 61

Keck, M. E., 155

Kehm, Patricia, 142

Kehm v. Proctor and Gamble, 139, 140

Kidd, Laura Klosterman, 137

Kilgore, Steve, 147–49

Kirbat, P., 168

kitchen, 38, 45, 188; waste, 184

Koothan, P. T., 162

Korea, 81n24, 147

Kotex, 140, 181
Kraditor, Aileen, 13

labor, 67, 185; labor-saving technologies, 119; market, 130; women's, 120, 184, 213n2
lactation, 56, 119, 122, 130, 132n4; advocacy group for, 120; lactating women, 127, 128, 129, 130; rooms for, 128, 129, 130, 131, 133n15, 209, 213, 213n3
La Leche League, 120, 122, 133n11
Lane, Brenda, 63
Larsson, Olle, 124
Latin America, 155
Latour, Bruno, 40, 207
Law, John, 39
lawsuits, 27n16, 133n12, 139, 140, 156, 193
lay expertise, 126, 174
Layne, Linda, ix-xii, 1-35, 55-118, 126, 179-96, 197, 201, 204, 211
Leadership in Energy and Environmental Design (LEED), 14
Leavitt, Sarah A., 89, 90, 94, 95, 98, 99
legislation, ix, 5, 11, 20, 21, 24, 81n25, 109, 126, 127, 131, 133n12, 138, 139-40, 142, 146, 150, 151, 160, 193, 205, 213n3
leisure, 5, 205
Leland, John, 114n33
lesbian, 13, 99
Lewin, Ellen, 13
liberation, 3, 23, 44, 57, 74, 136, 180; liberated, 136, 142, 150; liberatory, 4, 56, 66, 119, 124, 151; liberatory potential, 120, 130, 169, 173; women's, 4, 44, 83n43, 142, 144, 155, 180
Lockheed, Martin, 188
Logue, Zara, 191
long-term effects, 57, 167, 169
long-term use, 57, 77, 158
low-tech products, 109; lower-tech alternatives, 121. *See also* no-tech solution
Lunapad, 72, 82n40
Lybrel, 55

Mackenzie, Donald, 38, 120
Macklin, R., 167
Maclean, Sandy, 8
Maines, Rachel P., 13, 114n29
Malaysia, 66
male: culture, 62, 192; domination, x, 5, 62, 129, 194n7; fears, 150; male-centered norms, 199, 200; "male-friendly," 187, 188; male-valenced projects, 187-88, 194n7; methods of birth control, 7, 11, 28n19, 69-70, 77; privilege, 151
Maloney, Carolyn, 127, 133n15
mammograms, 119, 126
Manktelow, Nicole, 5
manufacturing, 18, 104, 137, 138, 140, 141, 146, 162; manufacturers, 4, 5, 27n13, 29n33, 96, 103, 111n5, 112n11, 120, 122, 124, 126, 144-46, 156, 160, 204-5, 211
market, 20, 24, 29n32, 55, 79n5, 80n16, 91, 93-94, 110, 111n5, 113n20, 140, 142, 144, 146, 149, 152, 156, 161, 172-73, 191, 198-99, 205, 206, 211; of breast pump, 124, 126; demands, 66; labor, 130; marketers, 103, 211; research, 138, 146; share, 22, 80n14, 104
marketing, 4, 5, 6, 7, 8, 17, 22, 26n9, 50, 56, 58, 64, 70, 79n4, 100, 113n23, 136, 139, 141, 142, 144, 208, 211
marketplace, 22, 80n14, 152, 204, 205, 206, 208, 209, 212, 213
Marshall, Robin, 104
Marso, Lorie Jo, 74
Martin, Emily, 70, 71
Martin, Joyce A., 101, 113-14n27
masculinity, 7, 69-70, 192; attitudes toward, 69; ejaculation-suppressing drugs, 69; masculinist, 43, 45; technology associated with, 37, 44
Massachusetts General Hospital's Institute of Health Professions, 61
Massachusetts Institute of Technology (MIT), 201
mass-produced, 136, 139, 185

materiality: change, 132; culture, 212; discourse, 152; of technology, xi, 37, 39, 40–41, 46, 47, 48, 49, 51, 154, 204, 212

materials, biodegradable, p. 145–46, 195–96; of products, 138, 139, 141, 185, 198, 200, 201

Maternal & Child Health, Bureau, 123; department, 114

Maternal and Fetal Health, 100–101

McDonagh, Deana, 24, 197–202

McGaw, Judith A., 3, 136

media, 24, 77, 159, 180, 186

medical devices, 141

medicalization, 158; of healthy women, 103–6; of pregnancy and birth, 110, 114n30

Melendez, Michele M., 5

Melzack, Ronald, 149

men, x, 6, 21, 27n17, 28n19, 28n23, 28n26, 69, 71, 77, 93, 99, 106, 107, 115n36, 152, 179, 180; attitudes towards menstruation, 62, 81n25; birth control for, 69–70, 77, 83n45; child care, 19; control over fertility, 168; equipment designed for, 5, 191; health, 81n24; housework, 19–20; median income of, 22; vis a vis women, 14, 28n23, 183; vs. women, 27n12; and women, 185, 198; world designed for, 137, 181

menopause, 18n21, 60, 68, 73, 80n21, 83n41, 83n43; peri, 77

menstrual cups, 72, 75, 143; reusable, 143; washable, 143

menstrual-disorders, 75

menstrual disturbances, 56, 58, 158–64

menstrual extraction, 75

menstrual suppression technologies, 55, 56, 79n6, 79n11, 119; birth control pills, 55–58, 104, 154, 207; pro menstrual suppression, 56

menstruation, 12; absence of, 158; as bodily fluids, 62, 128; as dirty, 68, 128, 141; as disabling, 83n41; as disgusting, 62, 82n37, 181; irregular, 77; late, 106–7, 114n29, 158; men's attitudes towards, 62, 81n25; menstrual-cycles, 104; missed, 99, 100; and moon, 68; as natural, 59; as necessary, 60, 61; as normal, 59; odor, 82n37, 138; painful, 7, 59, 67, 69, 73, 75, 82n37, 148–50, 209; periods-scheduling, 62; planned, 59, 62, 130; real, 59, 80n20; regular, 77; and sex, 62; as shameful, 82n37; as sign of womanhood, 67–68, 136; sponges, 59; spotty, 58; stigma, 58, 62, 113n21; taboo, 62, 81n24; as trademark, 67; as unnecessary, 60; women's unique biological processes, 136

Mexico, 81n24

Meyer, Donald, 18, 19, 77

microbicides, 24, 154, 170–74, 210

Midgley, A. R., 90

military: design for, 187, 194n7, 205; integrating women into, 5, 46

Miller, Leslie, 63

Mills, John Stuart, 205

Mintzes, B., 158

Mitchell, Robert, 128

Mitchison, N. A., 163

Mittag, Heinz, 142

modernity, 62–64, 68, 81n31, 119–21, 124–27, 130–31, 138, 205, 206, 208; premodernity, 81n31

Moellinger, Terry, 44

mortality: infant, 101; maternal, 101

Mother Earth, 68

Motherhood Lost, 8

Mother Nature, 59, 69

mothers, 68; blame, 114n28; as designers, 10; new, 8, 121; single, 13, 30n38; social opportunities of, 17; step-, 112n13; working, 119; young, 122

multidisciplinary, 189

Murkatz, Ruth, 70

Museum of Menstruation and Women's Health (MUM), 55, 58–76, 79n8, 83n41

MySpace, 50

Nash, H. A., 162
National Conference of State Legislatures, 213n3
nationalism, 66
National Public Radio, 19, 20, 22
National Science Foundation (NSF), xin5, 184–87, 194
Natracare, 144, 146, 200, 209
natural, x, 13, 56, 58–59, 77, 80n18, 80n20, 83n43, 113n25, 180; aging as, 60; antiseptic, x; birth, 63; choice of breastfeeding as, 126; differences between men & women, 180; drugs, 80n21; family planning methods, 59, 60; health care approach, 73; hormones, 90; menopause, 68; menstrual sponges, 59; menstruation as, 56, 58–59, 73, 79n8; resources, 143; unnatural, 80n18, 82n37
nature, 56, 58, 60, 63, 69, 71, 73, 74, 78, 80n21, 83n43; associated with women, 37, 68–69, 71; control over, 72; mimicking, 59–60; as mother, 59, 69; vs. culture, 74
Nature Medicine, 168
Nelkin, Dorthy, 97, 112n17
Nelson, Alondrea, 120
Nelson, Jennifer, 73
Neonatal Intensive Care Units (NICU), 189
Nestlé, 18–19, 121
New Delhi National Institute of Technology (NII), 162, 163, 166, 167, 168
New Mother's Breastfeeding Promotion Act, 132n15
New Zealand, 155
Nineteenth Century, 13, 137, 205
normalcy, 59, 74
Norplant, 18, 23, 73, 79n10, 83n41, 154, 156, 158; acceptability of, 163; disempowering, 154; problems with removal of, 158–59; safe removal, 159; safety of, 163
Norplant: Under Her Skin, 159
no-tech solution, 144

nursing, 8–10, 18–19, 120, 140, 209; emotional aspects of, 127; as "natural" choice, 126
NuvaRing, 79n6

Oakley, Deborah, 94
o.b., 142
Oblepias-Ramos, Lilia, 42
Oldenzeil, Ruth, 25n1
online user groups, 129
oppression, 70, 161, 166, 172, 180; forces of, 72
organic, 10, 144, 145, 188, 209; product, 146
Ormrod, Susan, 23, 45, 46, 120
osteoporosis, 60
Oths, Kathryn S., 101
Oudshoorn, Nelly, 7, 83n45, 84n45, 90, 92, 136, 154
Our Bodies, Ourselves, 91, 93, 107, 142
Our Hope Place, 8
ovulation, 59, 80n20; calendar, 5; kits, 104; predictor, 104, 105; prevention of, 56; research, 57; suppression of, 56, 207; timing of, 113n25
Oxfam, 21

packaging, 44, 66, 100, 111n9, 112n18, 113n24, 207
pads: breast, 144; disposable, 8, 138; heating, 69; reusable, 72, 75, 138, 143–44; sanitary, 8, 10, 26n10, 59, 72, 75, 82n40, 136, 138, 139, 140, 142, 146, 181, 183
pagan, 68
Page, Karen, 81n24
pain, 59, 141, 167, 212; managing, 150; menstrual, 7, 59, 67, 69, 73, 75, 82n37, 148–50, 209; perceptions of, 149, 212; of pregnancy loss, 15, 103
Pakistan, 81n24
Palmer, Gabrielle, 121
Park, Alice, 81n30
passing, 143, 181, 186

reviewers: of design work, 17, 184, 186, 190–91; of products, 126, 164; selection of, 190

Richardson, Frederick S., 139

Richter, Judith, 163, 164

rights, 13, 80n14, 94, 159, 169, 186; abortion, 155; disability, 2, 12, 193; individual, 156, 213n4; property, 205; reproductive, 73, 155, 156, 159, 166; of technology use, 6; vs. choice, 80n14; women's, 24, 51, 94, 127, 158, 166, 167, 168, 205, 210, 211

risks, 79n3, 107, 169, 171; acceptable, 163; of anti-hCG vaccines, 174; assessments, 169; and benefits, 105, 131, 167, 174; of cancer, 55; of cross-reactivity, 163, 164, 174; of family planning, 7; health, 7, 55, 57, 58, 89, 164, 165, 167, 169, 210; long-term, 169; of new products, 171; potential, 23, 103, 166, 174, 209; of pregnancy, 93; of public exposure, 97

rituals, 66, 68, 69, 81n25, 128

Roberts, Dorothy, 73, 81n23

Rock, John, 60

Rockefeller Institute for Medical Research, 56

Roe v. Wade, 193

Rosser, Sue, 82n36, 180, 186–87, 194n5

Rothschild, Joan, xin3, 2, 25n3

safety, 1, 5–6, 7, 13, 26n11, 28n20–22, 76, 141, 151, 155, 156, 159, 162, 163, 166, 170, 172, 210; long-term, 165, 168; of new reproductive technologies, 156; of product, 209; shoulder belts, 137; testing, 141

sales, 104, 105, 140, 141, 145; retail, 70, 126

Samsung, 5

Sanabria, Emilia, 11

Sanger, Margaret, 56

sanitary products, 57, 113n21; napkins, 140, 144, 181. *See also* pads; tampons

Sarewitz, Daniel, 195n11

Saul, Stephanie, 57, 61, 80n16

Scandinavia, 162

Scarry, Elaine, 212

Schrater, A. F., 164, 165, 167

Schroeder, Patricia Scott, 22

Schultz, Emily, 20

Schumacher, John, ix

Science and Technology Studies (STS), 89, 92, 115n43, 179, 180, 200

science, 161, 192, 208; "Greedy Sciences," 192; public understanding of, 174; of reproductive immunology, 162

Science, Technology, Engineering, and Mathematics (STEM), 187, 188; attracting women to, 188

scientists, 159, 163, 164, 165, 169, 170, 173; technoscientists, 210

scripts, 136, 137, 141, 142, 143, 150, 151, 206, 209, 210; cultural, 8; empowering, 161; gender, 154, 161, 165, 169, 170, 173, 195n11; introductory, 159, 161, 167, 169, 172, 173; re-scripting, 136–55, 206, 209, 210

Seager, Joni, 119, 130

Seasonale, 55–88; advocates for, 57; in the news, 57; opponents of, 57

Segal, Sheldon J., 56, 57, 59, 60, 80n12, 81n24, 81n31

self-knowledge, 106–7, 114n31, 115n35

semen, 69, 82n38

Sen, Gita, 78

sex, 13, 21, 38, 58, 60, 62, 64, 68, 69, 70, 82n37, 83n43, 127, 128, 149, 181; changing, 43, 44, 204; desexed, 150; discrimination, 21; and gender, 2, 41, 43, 44; hormones, 90; unisex, 12, 208; worker, 83

sexism, 11, 12, 21, 43, 45, 52n5, 62, 69, 77, 182, 183, 187; trans-sexism, 43

sexuality, 27n16, 82n37, 150, 155, 172; abuse, 109; pleasure, 70, 171; safe-sex, 170; sex-partners, 93; sexual activity, 64; sexual attractiveness, 62, 81n26; sexual dimorphism, 27n12, 208; sexualize, 81n26, 150; sexual vibrator, 209; sexual vigor, 81n22; taboos related to, 62, 81n24–25

sexually-transmitted diseases, 55, 170

shame, 132, 142, 150, 212

Shapiro, Laura, 121

SHARE, Pregnancy and Infant Loss Support, Inc., 111n2

Shaw, Gina, 60

Shervington, D. 167

Shew, M. L., 105

shopping, 5, 20, 22, 27n13, 98, 99, 199

side effects, 56, 70, 83n41, 149, 158, 159, 160, 162, 164; hormone-related, 166; medicalization of, 159; unexpected, 167

Sikkink, K. A., 155

Single Mothers by Choice (SMC), 13, 95

Sivulka, Juliann, 205

Sklar, Kathryn Kish, 13

Slate, 131

smoking, 4, 44, 55, 101, 113n26, 115n39

social, movement, 205; vs. individual, 156

social fix, 21. *See also* cultural fix

social justice, 22

social sciences, 180, 184

Society for Aeronautical Engineers, 188

Society for Automotive Engineers, 188

sociotechnical system, 39, 40, 41, 46, 47, 48, 49–51, 52n1, 130, 204, 207, 209, 210, 211, 213; feminist, 49–51

software, 51, 108, 109, 132, 159, 189. *See also* hardware

South Africa, 150

spaces, xi3, 16, 24, 29n32, 120, 129–30, 131, 174, 184, 192, 198, 209, 212; public, 11–12, 26, 28n29, 32, 126, 183, 209, 212; public vs. private, 12, 26n11, 119, 208–9; semi-private, 26n11; semipublic, 129

Speculae, 126, 200

speed, 113n23, 122

sperm, 69, 82n38, 93; banks, 13. *See also* semen

spouses, 93, 96, 107

stakeholders, 171, 185, 210

Stanley, Autumn, 37

State (The), 3, 155, 160, 210; abuse by, 164, 169

Steiner, Peter, ix, 50

Stemerding, B., 168

Stenger, Sharon, 8

sterilization, 73, 155, 162

Stewart, Cara, 71

Stim, E. M., 97

Stimpson, Catharine R., 2

Strausser, Susan, 72

strollers, 11–12, 17, 18

Stryker Orthopedics, 5

students, 15, 200; diversity among, 197; engineering, 195n9; as entrepreneurs, x; female, x, 1, 15 26n11, 189, 195n9; as instrument of change, x, 200

studio courses, x, 24, 179–202, 216

subaltern: alignment, 154–55, 157–61, 173; position, 204, 205

suffrage, 24

Swain, Heather, 106

Sweden, 101, 124, 144, 162, 167

Talwar G., 162, 163, 168

Tampax, 12, 136, 139–41, 148, 151

tampons, 12, 59, 67, 72, 77, 113n21, 119, 128, 136–53, 181, 209; as contraceptives, 137; ecofeminist, 133, 143–47; as medical device, 137, 141; Rely, 141, 142; vibrating, 147, 209

Tancredi, Laurence, 97, 112n17

Taylor, Janelle S., 96

teamwork, x, 24, 124, 183–201

technological fix, 21, 130, 133n17, 138, 151. *See also* cultural fixes; social fixes

technological literacy, 199

technological somnambulism, 110

technologies, appropriate, 110, 199, 204, 212; appropriated, 64, 66, 82n34, 92, 103, 104, 152, 206, 207; enhancement,

61–62; information, 6, 89, 108, 126, 146; labor-saving, 119. *See also* reproductive technologies

technology: abuses of, 73, 160, 161, 164–69, 173; acceptability of, 60, 154, 159, 167, 171–74, 210; access to, x, 20, 77, 110n1, 126, 129, 130, 159, 161, 200; contestation over, 154, 161–77; definitions of, 3, 25n1, 26n8, 181; development, 7, 23, 24, 38, 45, 50, 77, 78, 79n10, 131, 132n8, 140, 152, 154, 157, 163–74, 205, 211; emerging, 163, 198; failed, 56, 158, 203; future, 24, 56, 132n8, 154, 170; high-tech, 109, 168, 170, 172, 203; household, 185, 204–5; inadequate, 203; low-tech, 109, 121; materiality of, 204; military, 205; one-size-fits-all, 157; sexist, 21; "unintended consequences" of, 3, 11–12; valenced, 25n2, 76, 77, 195n11; wartime, 205

technosocial systems, 21, 24

teens, 29n31, 57, 58, 68, 77, 80n20, 92, 94, 98, 102, 105, 107, 112n13, 112n19, 113n22, 113n26, 122, 194n9

television, 28n23, 78, 95, 98, 99, 112n15, 144

temples, 16, 66

Tenderich, Gertrude S., 140

Thailand, 157, 158

Third World countries, 18, 42, 160, 165

Thomas, Sarah, 57, 58, 59, 60, 61, 66, 79n9, 80n12–13, 82n31, 104

toilets, 11, 27n16, 68, 140, 144, 183, 191, 198, 212; men and women, 195n10. *See also* bathrooms

touch, 140, 142, 200

Toxic Shock Syndrome (TSS), 141–42, 146

Tujak, Laura, 13

Tulane Xavier National Center of Excellence in Women's Health, 131

Turkey, 80n18, 80n21, 82n31

Twentieth Century, 25n1, 26n10, 103, 128, 137, 205

Twenty-first Century, 23, 26n10, 82n37

United Kingdom, 12, 17, 18, 21, 81n24, 81n27, 91, 99, 126, 144, 160. *See also* England

United Nations, 19, 160

U.S. Centers for Disease Control (CDC), 121, 141

U.S. Department of Energy, 188

U.S. Department of Health and Human Services, 101, 123

U.S. Department of Labor, 19

U.S. FIRST competition, 187

U.S. Food and Drug Administration (FDA), 55, 57, 91, 111n5, 122, 141

U.S. Market, 80n16, 94, 110, 113n20

U.S. National Institute of Health (NIH), 89, 90, 95, 97, 105, 107, 111n3, 112n19

University of Amsterdam, 155

University of Illinois at Urbana-Champaign, 197

University of Kansas Department of Obstetrics and Gynecology, 148

University of Michigan Institute for Social Research, 19

University of Oregon, xi, 15, 184, 191, 193

University of Pittsburgh's Magee-Women's Hospital, 60

University of Washington, 63

University of Wisconsin, 19

Unnayan Bikalper Nitinirdharoni Gobeshona (UBINIG) (Policy Research for Development Alternatives), 157, 158, 160

Upjohn, 56

urine-based pregnancy tests, 89–92, 94, 104, 106, 111n3, 111n5, 113n23, 115n37

Uruguay, 89, 101, 111n9

user-centered design, 24, 131, 152, 154, 164, 168, 170, 174, 199

user manuals, 93, 94, 106, 111n7, 151, 181; instructions, 108–9

users, 92, 126, 171, 180, 204, 206; actual, 169; benefits to, 166; category of, 24; current, 168; diversity among, 170, 173;

experiences, 171; failure, 165; future, 154, 157, 167, 173, 174; infrequent, 183; male, 168; needs projected by, 169; potential, 149, 156, 168, 206; presumed, 92, 137, 154; ultimate, 210; women, 20, 24, 161, 168, 208, 210, 211, 214n5

Vaitukaitis, Judith L., 90, 91
Valence, feminist/antifeminist, 76, 77; male, 187, 188, 194n7; of technologies, 25n2, 195n11
vehicles, 145, 187, 188, 189, 194n3
Vestby, Guri Mette, 52n1
vibrators, 13, 25n2, 147, 149, 209
violence: against women, 5, 7, 13, 26n11, 28nn20–22, 209; domestic, 7, 28n20, 212. *See also* rape
Violi, Paula, 9
Vipon, 75, 147–50
Virginia Slims, 4, 44
visibility, 124–27, 181, 183, 185. *See also* concealment; invisibility
Viswanath, K., 168
Vostral, Sharra L., x, 7, 12, 24, 26, 75, 80n17, 113n21, 124, 126, 128, 136–53, 181, 197–202

wage, 21, 22, 180; compression, 121; equal, 211; gap, 20; high-wage jobs, 121, 130; living, 146; lost, 61; low-wage jobs, 130; minimum, 146; women's, 20; work, 19–20, 21, 112n14, 119–33, 146, 209; workplace, 119, 120, 127, 128, 211
Wagner, David, 11
Wajcman, Judith, 11, 25n3, 25n5, 26n9, 37, 38, 120, 152
Waldby, Catherine, 128
Walker, James R., 7
waste, 82n40, 98, 105, 143, 144, 146, 184, 201. *See also* disposal
Weber, Rachel N., 46
Webster, Juliet, 120
Weigl, Andrea, 7

Western Europe, 91
Westhoff, Carolyn, 59
WGBH Public Broadcasting Station, 194n9
Wheeler, James, 131
Wide, L., 90
Wi-fi, use by women, 19–20
Wilcox, Allen J., 93, 113n25
Winner, Langdon, ix, 48, 110, 179, 193, 194n2, 195n11, 198, 206
Winston, Carla A., 101
Womanly Art of Breastfeeding, The, 122
women, African-American, 72–3 122–23; consumers, 22, 89, 137, 141, 147, 199; designers, x, 8, 28n25, 154, 181, 188, 200; with disabilities, 66, 79n9, 194n1; diversity among, 3, 18, 20, 24, 26n10, 55, 58, 68–69, 72, 78, 97, 124, 130, 152, 166, 173, 179, 180; elderly, 20, 194n1; in engineering, 188–89, 194n8; entrepreneurs, x, 7–10, 28, 28n25, 144, 198; leaders, x, 189; low-income, 21, 77, 98, 101, 122, 130, 211; married, 20, 99; middle-class, 20, 64, 78, 79n8, 101, 130; minority, 101, 113n27; modern, 27n13; in politics, 21, 22, 127; in the professions, x, 5–6, 95, 120, 188; "real," 67; in relation to men, x, 6, 13–14, 28n23, 67, 128, 132; single, 13, 30n38, 95, 98, 99; teenaged, 1–9, 29n31, 58, 67, 68, 80n17, 92, 95, 98, 102, 105, 106, 107, 112n13, 112n19, 113n22, 113n26, 122, 194n9; upper-class, 64, 78; users, 24, 167, 198, 208, 210, 211, 214n5; vis a vis men, 61, 180, 181, 183; working, 27n13, 27n16, 61, 64, 149; young, 58, 64, 68, 81, 83, 95, 102, 111n8, 113n22, 138, 187, 188, 191, 192
woman-headed households, 22
women and technology studies, 2
women-centered design pedagogy, 198
Women Engineers Project, 188; Extraordinary Women Engineers Project, 88–89, 192–93, 195

women-friendly objects, 200, 201
Women of Color Reproductive Health
    Forum, 167
Women's Design Service, 20, 21
Women's Global Network on Reproduc
    tive Rights (WGNRR), 155, 164–69
women's groups, 20, 21, 160
women's health, 146, 157, 158; advocates,
    24, 110, 154, 155, 161, 170, 171; centers,
    75, 91, 106; movement, 26n10, 56, 76,
    79n10, 89, 141, 142, 155, 156, 166, 169;
    organizations, 155, 161, 210, 214n5;
    publicly-funded, 109, 157
Women's Health Advocates on Microbi-
    cides (WHAM), 170
Women's Liberation, 3, 4, 155
women's rights, 158; movement, 205
Woodhouse, Edward, 22, 195n11
work, 19–21, 61, 70; absenteeism, 131;
    equal, 211; in the home, xiin7, 5, 19–20,

184–86, 204; meaningful, 188; outside
    the home, xiin7, 18, 62, 64, 71, 119–33,
    204; parity, 199; work-life balance, 22,
    63, 119, 131, 192, 193
workplace, 52n3, 61, 70, 120, 130, 183, 192;
    preindustrial, 205; safety, 52n3
World Health Organization (WHO), 24,
    80n11, 81n24, 123, 156, 159, 162–68
World Population Plan of Action
    (WPPA), 156
war, x, 138, 180, 205

Yanco, J., 166
Yankauer, A., 91
Yearley, Steven, 40
Yeastie Girls, 150
Young, Linda, 20

Zinamen, Michael, 126
Zonite Products Corporation, 140

Women, Gender, and Technology
Edited by Mary Frank Fox, Deborah G. Johnson,
and Sue V. Rosser

Feminist Technology
Edited by Linda L. Layne, Sharra L. Vostral,
and Kate Boyer

*The University of Illinois Press
is a founding member of the
Association of American University Presses.*

Composed in 9.75/13.25 Adobe Minion Pro
by Celia Shapland
at the University of Illinois Press
Manufactured by Thomson-Shore, Inc.

University of Illinois Press
1325 South Oak Street
Champaign, IL 61820-6903
www.press.uillinois.edu

represents a way to deliver even more fully on the breast pump's promise as a feminist technology.

BODIES OUT OF PLACE

We have argued that breast pumps can be considered a feminist technology in that they expand both mobility and choice in the field of infant nutrition. We now turn to what happens when users attempt to deploy this technology in real-world situations, particularly in the wage workplace. Although breast pumps can enable lactating women to return to work, actually pumping at one's place of employment can be difficult. At the most practical level, pumping at work is not a guaranteed right in every state, and women have been fired for it (Blum 1993, 291).[9] New York Congresswoman Carolyn Maloney noted as a reason for introducing the Pregnancy Discrimination Act Amendments of 2000, designed to protect woman's right to pump at work: "Women who choose to breastfeed have no choice about pumping milk during the day; they simply must express milk regularly." Maloney continued: "When women have stood their ground and told unrelenting bosses that, like it or not, they need [break time] for pumping, these same women have had their pay and benefits docked, and even lost their jobs."[10]

Indeed, even the breast-feeding advocacy community has been ambivalent about using breast pumps as a means to return to work. As Hausman has noted, there is a tendency within this community to view the breast pump as a technology that diminishes the emotional and psychological aspects of nursing (Hausman 2003), and historically this community's stance has been that women with young children should stay at home and nurse as long as they can (Bobel 2001). This position might be interpreted as only qualified support for women seeking to combine work and nursing, or possibly an anti-technology, anticonsumerist preference for manual expression, as discussed earlier.[11]

Even where pumping in the workplace is legal, problems can remain both at the level of built form and cultural practice, in that most workplaces "design out" nearly all activities other than work itself; for example, worksites in which one cannot eat or sit down (such as most stores) and in which bathrooms are too few (such as outside and/or male-dominated worksites) show how workplaces sometimes deny the physical needs of the body for employees of both sexes. Experiences of women seeking to pump and store milk at work, however, offer an illustration of how breast-pumping women—and their bodily products—are constructed as particularly "out of place" at work.

The lactating body in the workplace causes anxiety both because it draws attention to women's biological productivity and because it involves the purposeful excretion of a bodily fluid. As Boswell-Penc has argued, breast milk is sometimes viewed with suspicion as a potentially contaminated and contaminating substance (Boswell-Penc 2006). This builds on Mary Douglas's argument in *Purity and Danger,* which states that across many cultures, bodily fluids have been understood as dirt or pollution because they traverse what is usually the firm boundary of the body, thereby functioning as a symbol of danger, disorder, and power. Transgressions of the body boundary are fraught with ritual and particular social codes and, in many cultures, the bodily fluids of one sex are thought to pose a particular threat to the other (Douglas 1966, 120–21). Yet along side this disparaging view of breast milk stands an opposite interpretation: that of breast milk as "liquid gold," thus considered for its unique immunological and hormonal benefits that science has yet to find a way to replicate. In this respect, breast milk perhaps bears something in common with the way Waldby and Mitchell argue other "mobile" bio-objects are now viewed, as alternately precious or threatening depending on the circumstances (Waldby and Mitchell 2006).

We can find contemporary references to the fear breast milk can illicit in the 2003 case of an Albany, New York, woman who was fired for storing her breast milk in a communal refrigerator at her work.[12] This kind of anxiety also echoes old narratives of fear about women's entrance into the white-collar workplace in the early twentieth century, which produced impressive efforts to physically separate men and women employees and provide for women's "special needs" (a code word for menstruation) such as by providing spacious anterooms to bathrooms complete with sofas that could be used for rests (Boyer 1998). Such efforts can be interpreted variously. In one reading, such interventions reflect an arguably regressive desire to contain or quarantine women's bodies and their effluvia from the rest of the workplace. However, if we instead view these early spatial interventions as being of a piece with the provision of lactation rooms in today's offices, an alternative reading is that such accommodations instead constitute reasonable—indeed progressive—efforts to acknowledge and accommodate women's particular biological realities within the wage workplace.

As opposed to technologies that hide uniquely female bodily functions (such as tampons) that, as Vostral argues, are designed to *un*mark women's bodies (Vostral, this volume), the breast pump calls attention to a uniquely female bodily function, sometimes in a fairly dramatic way. Electric pumps, which were favored by nearly all of our interviewees for their efficiency, are also quite loud, and this was a source of anxiety for lactating women who chose to pump at work. As noted, in both pump advertisements and Web sites devoted to breast pumping, the noise level of breast pump motors is identified as problematic or embarrassing. For example, as Dilys Wynn, one nursing mother interviewed in the article "Express

Yourself: How to Successfully Combine Breastfeeding and Work," explained her experience: "I didn't tell my manager or my co-workers that I was expressing milk at work. . . . It was a male-dominated industry and it would have been too embarrassing." To mitigate her embarrassment, Wynn pumped in her workplace medical office.[13] One woman in our sample used a manual pump at work as a way to reduce noise, but said that she got tennis elbow as a result.

Pumping requires both time as well as a place that ideally is private, calm, and sanitary. Pumping has to be arranged so as to correspond roughly to the frequency of nursing. Because nursing is a case of supply meeting demand, lactating women need to express regularly by one means or another in order to continue to produce milk. Our research suggests that this amounts to breaks of about fifteen to twenty-five minutes every few hours. In pumping advertisements and in online user groups, the need for privacy is also identified as being essential to the pumping process. Our interviews reveal a variety of strategies to achieve what they viewed as the "appropriate" degree of isolation. None of the women we interviewed had access to dedicated lactation rooms. Some pumped in their cars, some in bathrooms, and some found small unused rooms or closets. One interviewee told of a colleague who worked in state government who wound herself up in a long window curtain. Anxiety about discharging a bodily substance in the workplace or drawing attention to one's breasts by having them manipulated by a loud machine within earshot of ones' colleagues, together with the desire to avoid adding another layer of spatial and temporal discipline on top of those already required by one's job, may all serve as disincentives to pumping at work.

We further submit that the emphasis on concealment and banishment described in our interviews suggest that lactating women almost certainly feel more embarrassed to be seen by others engaged in pumping than to be seen nursing. We did not ask about this directly, but it squares with our broader experiences. We have each witnessed colleagues breast-feeding in semipublic spaces of conference rooms, classrooms, faculty meetings, stores, and restaurants. In 2009 in upstate New York, it is very difficult to imagine someone pumping in any of the aforementioned situations or spaces.

TIME, SPACE, AND CLASS

Further, bracketing any inconvenience or embarrassment pumping may produce, being able to engage in this activity in the first place depends on being able to secure the requisite time and space. Here, the onus is on employees (rather than employers) to find a way to make pumping work. As Dilys Wynn put it: "It is up to you to work out where to pump, where to chill and store your milk and how to schedule work breaks that coincide with let-down times."[14] Comments such as these suggest that it is up to individual employees to convince one's employer to allow them to pump at work. Most workplaces do not provide additional breaks to lactating women; thus, as Haidee Allerton points out in her article entitled "Coffee . . . Uh, Milk . . . Break," many women report having to accomplish their pumping during lunch or coffee breaks, around the edges of their workday (Allerton 1997).

Presumably, savvy employers in highly remunerative sectors of the labor market who value their female employees' time and want them back at work as quickly as possible after childbirth should provide time and space to allow their employees to blend nursing or pumping with wage work. And indeed, as feminist geographers Mona Domosh and Joni Seager note, one of the leaders in workplace lactation practices has been financial company J. P. Morgan Chase, which provides a lactation room with built-in pumps on their trading floor (Domosh and Seager 2001). Of course, most women do not work at J. P. Morgan Chase. And as a corollary to the gusto with which some high-wage employers have embraced workplace lactation, we suggest that finding time and space to pump is especially difficult in low-wage jobs. Space costs, so women who work in fast-food restaurants, coffee shops, or mall stores are very unlikely to have access to the kind of "extra spaces" such as dedicated lactation rooms, examination rooms, or empty conference rooms that middle-class women might use. Thus, whether one chooses to continue nursing after returning to work is not only a question of cultural preferences or the cost of a breast pump; it is also an issue of workplace design and whether one works for an employer who will make space for this activity.

CONCLUSION

Breast pumps, then, function as part of a broader sociotechnical system that includes workplace design and the social politics of actually pumping. Although we argue that the breast pump is emancipatory in that it expands mobility for some women, we also suggest that current social factors in the United States constrain this technology's liberatory potential. We are concerned that the "goods" of breast pumping are distributed unevenly to women employed in workplaces that will make room for pumping, and we postulate that one reason low-income women do not continue nursing as long as their middle-class counterparts is because they are less likely to work in such environments. At the same time, we are also concerned that by providing a personalized, technical fix to the question of workplace lactation, breast pumps could inadvertently remove the incentive for employers